KU-614-053

MONTY DON

Gardening at Longmeadow

Photography by Marsha Arnold

Contents

Introduction 7

January 12
February 28
March 46
April 76
May 128
June 178
July 220
August 264
September 312
October 356
November 394
December 418

Map of Longmeadow 440
Index 441
Acknowledgements 448

Introduction

I first saw this garden on a particularly dank autumnal day in 1991. The front was covered with piles of building rubble. At the back was a little yard filled with knee-high weeds happily seeding themselves, and beyond that was a paddock where a grumpy horse tried to find sustenance amongst the brambles. There was nothing here that could possibly be described as a garden. But beneath the years of neglect was a blank canvas that I could fill with the garden of my dreams.

For the first six months all my time and energy was directed towards the house – which was an uninhabitable ruin – and it was not until the following spring that I began clearing the field at the back. Three times that summer I cut the grass, and three times I raked it up along with buried and tangled tree trunks and discarded farm machinery. To get to know the land through hard graft was better than through a drawing board. In fact, this slow gestation was the best thing that could have happened and I recommend it to anyone making a garden from scratch. Take your time. Make and unmake it in your mind until you are ready to begin. You will know in your bones when that time comes.

Then I persuaded a neighbouring farmer to plough the cleared land for me. The turf unfurled to reveal rich, dark soil. Everything grows lustily in it. Even the weeds – especially the weeds – are astonishing in size, vigour and range. That vigour is a huge advantage for any gardener. Mind you, this was just about the only apparent advantage. Otherwise the odds were stacked against me. The site was wind-blasted and needed shelter. To all horticultural intents and purposes, it was empty: there was just one large hazel outside the back door and a hawthorn at the edge of a ditch that bisected the field.

Then I had my big break. On the famously chaotic day of the Grand National in 1993 – April 3rd – I went to a local tree sale with a proposed budget of £200 intending to buy some good-sized yew plants for a hedge. I came back five hours later having spent £1,300 on 1,400 trees and hedging plants. It poured with rain all day and by lunchtime the allure of the pub and the Grand National was enough for most people to leave the sale. A tiny handful of us stuck in, soaked and buying increasingly large lots at increasingly minuscule prices. The last batch of 15ft-tall *Tilia platyphyllos* I bought – and which now make up the Lime Walk – were 50 pence each. However, only a frantic phone call to the bank – in the days when managers were real people – increased our overdraft to cover the cheque. But this was the critical moment that made this garden.

I also bought a batch of very cheap box plants that I had learnt about through the local paper. When I went to collect them, they turned out to be an established and neatly clipped hedge. I dug it up and replanted it as two parallel hedges in the Ornamental Vegetable Garden, where it has remained for the past 18 years and provided thousands of cuttings. (The Ornamental Vegetable Garden has also provided us with thousands of meals from its rich Herefordshire loam.)

Although I had played it all out in my head before we began planting, there has been quite a lot of trial and error at Longmeadow. I never think of it as finished – just where it happens to be now. We have moved trees and even entire hedges and we are constantly replanting borders. I am a great believer in moving plants until they are absolutely at home, and I do it all the time. We made mistakes too, and I wish that most of the paths were wider. We planted the main hornbeam hedges in 1995 and although I knew how high I wanted them, I had underestimated how wide they would grow. We are constantly reducing their width.

Longmeadow is a garden centred in its geographical place, which is the Herefordshire Marches, just eight miles (as the crow flies) from the Welsh border. It is dead flat and hard by a river, and about a third of the garden regularly floods. The soil is clay loam over gravel, which is wonderful when dry but intractable mud when wet and rather heavy and slow to warm up. It is a very cold site, exposed to the wind and the rainfall is very high. But the western wind quickly blows away the bad weather and the rain means that we rarely suffer from drought.

The garden is essentially a rectangle with the house in one corner. This made the design awkward so I made a path across the width of the site coming from the main door that leads from the house with three longer paths leading off at right angles through the length of the garden. Other paths cut across these three main highways to create an irregular grid that the garden and all its 19 different parts have come to fill.

These different sections (that have gradually acquired names – and separate identities – over the past 20 years) include a small Spring Garden just outside the back door, an Ornamental Vegetable Garden which is a formal grid of box-hedged beds filled with colour, exuberance and good food within a tight structure. Come to think of it, that probably describes my garden at its best. There is also a more straightforwardly practical vegetable plot that we call the Top Veg Garden – not because it is necessarily the better of the two but because it is at the top of the garden – an Orchard and a Soft Fruit Garden.

The Jewel Garden is the largest area and right at the centre of the plot. It is filled with only jewel- or metallic-coloured plants for maximum impact from spring to autumn. The Walled Garden is in front of the house and surrounded by a stone wall on two sides, and that yew hedge I went to buy plants for on the fateful Grand National day in 1993. We have a Coppice filled with flowers in spring; a Damp Garden that is the first to flood; and a Dry Garden in the front made on almost solid stone.

Some parts feel like rooms at the end of a corridor that you have to make a special journey to visit while others are communal spaces or even the corridors themselves. But I think it all hangs together and, most importantly, has become irreducibly itself and is more than the sum of its horticultural parts.

It is where I garden – and although I have written millions of words about gardening and made television programmes about it for a quarter of a century – I never think of gardening as an objective process. It is what I do in real life in my real garden. It is a record of failure, bewilderment and surprise, as well as endless pleasure and some success. It is about life in all its complexity, sadness and joy as much as the intricacies of horticultural technique.

I do believe that most good gardens are personal, private, domestic and above all, intimate. The measure of their goodness can be reckoned to some degree in absolute terms of design, planting or horticulture but not to any meaningful extent. The truth is that our response to gardens is invariably subjective and if they are our own, completely so. To an astonishing and powerful degree they are loved, and love cannot be reasoned or measured.

Which puts me in rather an odd position. Over the years I have often written about this garden but have never been answerable to anyone or anything other than private whims and fancies. It was made as a wholly domestic place where, over the past 20 years, my family has grown up. The other critical point is that it is not mine alone and nor would I want it to be. Every tiny detail is shared and it has always been a joint venture with my wife Sarah. Nothing in it has happened or been planted without discussion and both of us have an absolute power of veto over the other. In every sense it is as much her creation as my own and that fact enriches my own enjoyment of it.

Yet now it is shared with more than two million people every week via the medium of *Gardeners' World.* The very private has spilled into the very public. Although I have made gardening television continuously since 1988 (including a five-year stint with *Gardeners' World* at Berryfields from 2003–2008) the combination of professional and private in one's own back yard is very different. But that potentially awkward balance is exactly the reason I agreed to do it. The challenge was irresistible.

I can not completely separate my passion for gardening from my passion for this garden. This is probably a flaw – but although other gardens can be visited, admired and analysed, they could never replace the depth of involvement that I get daily from my own plot. So when the opportunity came up – completely out of the blue – I took a deep breath and realised that I had to do it.

So, although much of the horticulture in this book is based upon received wisdom or knowledge and techniques shared by many generations of gardeners, it is personal. Everything in it is based upon our experiences of gardening here and how we go about making and tending it across the days.

Although I share a lifetime of experience and knowledge I make no apologies for idiosyncrasies and particularities. Technique and skills are important and useful but only as a means to an end, and the only end that matters a jot in gardening is to make a beautiful garden that will give you sustenance and pleasure and to be able to share them with those you love. Everything else is secondary.

The lollipop globes of **Allium hollandicum** *'Purple Sensation' are composed of scores of individual florets creating an open carapace of flowers that not only look dramatically stylish but are excellent for wildlife.*

JANUARY

Every year I have an almost tangible sense of renewal in January. Part of it is sheer relief. November and December are my two least favourite months and I am glad to see the back of them.

This is mainly to do with light. I hate the darkness of a British winter. For me it is less of a cosy time to hole up in front of a fire than a dark, dank tunnel that I have to crawl through in order to reach the light at the other end. And now, at the end of the year, I can just see a tiny pinprick of hope at the end of that tunnel. Every day is stretching out, just little by little and that is enough motivation to keep me moving towards the light.

Now I am sure that this is a gardening thing. Many people I know get most down in January and February, finding that after the New Year the winter stretches out ahead of them. But if you are a gardener then these next few months are an exciting, increasingly busy time. The garden starts to peek its head up from below the ground. Snowdrops, aconites, hellebores, winter honeysuckle, mahonia and viburnum are all pushing into flower.

The structure of Longmeadow comes into its own and on a crisp frosty day the garden is etched in clean, strong lines. It is stark but strong and the balance and proportions are very pleasing.

Then there are all the things I ought to do and have left undone. Leaves to gather, ground to prepare, garlic to get in the ground as soon as possible, onion seeds to sow and sets to buy – in fact all my vegetable seeds to order – the greenhouse to wash, tools to mend and go through, and the potting shed to give a thorough tidy out. These are jobs that should be done at the end of the year but if I am honest I never do them then. Thinking about it just makes me feel tired. Now it feels like tidying the kitchen in order to make a lovely meal, whereas at the back end of the year these jobs feel like clearing up after someone else's mess.

In January, minute-by-minute, the days lengthen and hope creeps back into my world. I cannot tell you the relief. There is a hawthorn in the boundary hedge of Longmeadow. It is a scrubby affair, not much more than a bush really, but every mid-January the sun lingers just over the top of it before dipping down over the horizon across the fields. This is an important day because that light shines right down the garden and catches the panes of my greenhouses. The garden is literally lit up for the first time since October.

ACONITES

The buttercup yellow flowers of winter aconites (*Eranthis hyemalis*) are usually the first to open in January and are the brilliant midwinter counterpoint to snowdrops' modest charm. Their flowers, fringed with a green ruff, open in the winter sunshine, reflecting light, then close again at dusk.

They are bulbs – or more accurately rhizomatous tubers – but spread by seed very easily once established. When the plants flower they are without a stalk but this develops to carry the seedhead, raising it above the surrounding fallen leaves and grass so that the seeds can scatter better. One way to promote a good spread of the plants is to strim the ripe seedheads, flinging them further than their natural distribution. 'Guinea Gold' flowers a little later than the common aconite with bright orange flowers and a bronzed fringe, or involucre, creating a dramatic contrast. It prefers more shade than the common version.

It is best to plant them 'in the green' which means as plants just after flowering. They like damp shade and the base of deciduous trees is ideal, but because the flowers only open up in sunshine they do need some sun during the short winter daylight hours. Mind you, the sunshine can be accompanied by frost and icy snow and still the flowers will open, which seems to me to be as good a reason as any for getting out of bed on a winter's morning.

It is important to plant the rhizomes at the right depth, which is generally rather deeper than one might think, with the top of the roots about 8–10cm (3–4in) below the soil. They prefer an alkaline soil and good drainage and plenty of organic matter in the ground – which, of course, they would get from leaf mould in their natural habitat.

SNOWDROPS

I planted the snowdrops in the Spring Garden 15 years ago now. The first batch were a present lifted from a friend's garden and delivered wrapped in damp newspaper, and they have gradually been spreading by seed – at about the rate of 2.5cm (1in) a year – although every few years I do lift and divide a clump or two. Left to their own devices they will gradually carpet the entire area they occupy, with the rich, rather damp soil that they love, and some shade that also suits them. The pollination of snowdrop seed depends upon two things, some sunny, mild weather and the insects to spread the pollen. In the case of snowdrops the outer petals open to be horizontal when the temperature rises to about 10°C (50°F) and this attracts insects. The green markings on the inner petals (that every snowdrop has to a greater or lesser extent) are said to glow in ultraviolet light, which is another enticement for pollinators like the queen bumblebees that one sees bumbling around in the winter sun.

In fact, the best way to make a clump of snowdrops spread is to lift them immediately after flowering, divide up the mass of bulbs and replant them in smaller groups a few feet apart. Seed dispersal will mean that these clumps gradually meet, and repeated division every few years will further speed the process greatly.

Unless they are growing in grass then you are almost certain to disturb them when and if you plant anything else near to them – which you are almost certain to do as they disappear to nothing by midsummer and do not amount to much after mid-spring, as their foliage gradually withers. (As with all bulbs, resist any temptation to tidy or cut off that foliage because every last scrap of green is essential for the formation of a healthy bulb for next year's flowers.) In my experience, snowdrops are pretty good at dealing with the trauma of the occasional excavation and as long as they are popped back into the ground quickly they do not seem to suffer. However, a way round this risk is to plant them around the base and roots of deciduous trees and shrubs where they are less likely to be disturbed and will not mind the summer shade. The only thing to watch for is the ground getting too dry – especially in autumn when they start to grow again, albeit underground and out of sight for another few months.

Snowdrops are good as a cutflower if you pick them with a longish stalk. The first tiny bunch of modestly inclined flowers in a vase on the kitchen table is a wonderfully hopeful moment and they have a surprisingly strong honeyed fragrance drawn out by the heat of a room. They also grow well in pots and are ideal for small terracotta pots that you can sometimes find in large quantities at car boot sales. Use a general-purpose potting compost to plant a small clump in each pot and keep them outside

in a cool corner, bringing them into the sun in the New Year. You can bring the pots indoors to make a lovely houseplant, although the flowers will last longer outside in the cool.

They will not need repotting or feeding every year, but keep them watered from October through to June and every three or four years take them out of the pot, divide them into three, repot into fresh compost and let them get on with it.

No one seems to know if snowdrops are native or not, although because of their longevity and 'naturalness' they feel as though they ought to be. There is no reference to snowdrops growing wild before 1770, and indeed, the first garden reference is not until less than 200 years before that, in 1597. And although they seem carelessly natural they have been bred as intensively as almost any garden flower. There are over 350 different species and cultivars, although the differences are really very particular. The common *Galanthus nivalis* will do me fine although I do love the double *G. n.* f. *pleniflorus* 'Flore Pleno'. This is sterile, so will not spread from seed, but increases perfectly well from divisions and because it does not produce seed it has the bonus that the flowers last an extra long time.

Snowdrops in January sun. All our snowdrops are planted in the Spring Garden in a growing ribbon either side of the path and have spread from one small original clump.

CAVALO NERO

Cavalo nero or black Tuscan kale is the most useful brassica growing in my garden and although it is at its best in winter when the leaves have had some frost on them, I grow it all the year round. It can be eaten raw in salads when the plants are young, or left standing from summer through to the following spring for use as cooked leaves. Unlike most cabbage the leaves can be cooked for a long time so are great in stews, soups and sauces. As a pasta sauce with garlic and hot cream it is fabulous.

You pick the leaves individually and the plant replaces them with more and more fresh ones until it starts to flower almost a year after planting.

I sow the first seeds of the year in a seed tray or plugs in January with a couple of extra sowings at monthly intervals. If germinated in seed trays the seedlings must be pricked out into pots before planting out into their final positions in ground that has previously grown a leguminous crop such as peas or beans. The plants grow fairly large so need 60cm (2ft) in each direction and may need staking. If grown as a salad crop they can be sown directly into the soil in rows and thinned to just 10cm (4in). They are a brassica so will need protection from cabbage white butterflies between June and September.

LEEKS

Leeks are one of the few vegetables to stand happily through all kinds of winter weather. I have dug them in pouring rain and mud and when the soil has been frozen so hard as to be almost impenetrable. Yes, they freeze solid, so you bring them indoors as a mad, mud-encrusted, Heston Blumenthal-inspired lolly, but once thawed they can be eaten in all their glory.

When I was a child, leek in white sauce was my idea of a treat. That comforting, slightly slimy, bland but distinctive texture and flavour was my perfect comfort food. Still is, although my range of tastes has expanded a little from a Britain in the psychological – if not practical – grip of rationing.

I start sowing my first leeks in February, the wispy green hairs of the new seedlings sharing garden space with last year's crop for a couple of months. I will

Cavolo nero growing in the Top Veg Garden. It is are extremely hardy but delicious. Although good all year round, it is at its best after a frost.

make at least three sowings to keep a year-round supply. I used to sow in seed trays and then prick out into 8cm (3in) pots but nowadays I sow direct into the pots, a pinch of seed in each. I also used to plant the seedlings out (in May – as the residues of last year's crop are making wonderful minaret flowerheads) individually into holes made with a dibber, made in turn, from the handle of an old spade. It was how I was taught to do it as a child. But I now plant in small clumps of 4–8 plants and use a trowel. Less ritual, less rhythm and perhaps less magic, but just as good leeks – and the clumps of small- to medium-sized stems are ideal for each meal. Big leeks, let alone giant ones, are absurd. Keep them small and sweet for the kitchen.

Leek rust (*Puccinia allii*) has been a problem over the last few years, thanks to the warm, wet climatically changed weather, and I guess will continue to be so. Wider spacing, less compost and even tougher hardening-off regimes will encourage less soft growth which will help, but not stop, the problem.

I am fickle with my loyalty to varieties, although I try to grow at least one heritage variety each year. 'Musselburgh' is the oldest British variety, 'Pandora' has a blue tinge to the leaves and is fairly rust-resistant, and 'Varna' is one of the best for mini leeks, which, my biodynamic, market-gardener friend tells me, is the best-selling vegetable they grow.

ONIONS AND SHALLOTS

I sow a batch of onions in the New Year to give them the longest possible growing season, although they will need the protection of a greenhouse or coldframe for another few months, and then a period of hardening off before they can be planted out. I take huge encouragement from this first creative act in the garden of the year, starting new life with the promise of a summer harvest created from this point in the depth of winter.

The onion is one of the oldest vegetables cultivated by man and references to onions are found dating from 3200BC in Egypt. One of the reasons why people have always grown them is that they are obligingly easy to get right.

There are two ways of growing them, either from sets or seed. Sets take about 20 weeks to mature from planting. Seed takes perhaps another four weeks on top of that. There are many more varieties of onion available from seed but sets (which are just small onions) are easy to grow and so I always grow some as well as seed. You buy a bag of the small bulbs of a variety that appeals, prepare the soil so that

PLANTING ONION SETS

I fill a tray of plugs with seed compost and carefully insert one onion or shallot set into each one, dibbing a hole so the root plate is not damaged.

After watering thoroughly I place the trays on a bench in the greenhouse. They are very hardy so do not need any extra heat or protection.

After about a week green shoots will appear from the tip of each set and this indicates that roots are growing into the compost.

When each set has a couple of inches of healthy growth, I gradually harden them off outside and transplant them to their growing position as soon as the soil is workable.

it is fine and soft enough to stick a finger in to the knuckle without any soil sticking to it and then bury the sets so that the tops are sticking out of the ground.

I use a scaffolding board as both a straight edge for the rows and for me to kneel on to avoid compacting the soil. It is a good idea to make sure the sets are in a true grid because then you can hoe in both directions without clipping a bulb in passing. It is important to hoe – and occasionally hand-weed – onions as they respond badly to competition for water and nutrients.

Onions do best in good but lightish soil. If the ground holds too much manure or compost you will have lots of lush leaves but the onions themselves will be on the small side and more prone to fungal problems. Size does not matter so much – in fact a good batch of medium and small onions is more use than ones the size of croquet balls.

I think shallots are as important as onions and they are easier to grow. They tolerate poorer soil, hotter and colder weather and less water than onions. They also have a distinct sweetness of taste and they store much better than onions.

The real difference between onions and shallots is that each individual set or seed will multiply and produce a clump of around half a dozen small bulbs. These are harvested in exactly the same way as onions and when dry I store ours in a wire basket in the potting shed. When we need some for the kitchen I just go and scoop up a handful. If they are kept cool and dark they should reliably store well into spring.

The one essential is that they are as dry as possible before they are stored and the more sun that they have before harvesting the better they will last. As with all bulbs, it is important not to cut off any of the leaves but let them die back completely and dry out before removing the residue for storage. Onions should be lifted carefully with a fork rather than yanked from the soil to avoid damaging the root collar and thus reduce the risk of rot entering the bulb when stored. I always harvest mine on the morning of a hot dry day and leave them to dry on the ground for the rest of the day. I then put them in the greenhouse or on a wooden rack for a few weeks. When they seem to be bone dry I either plait them and hang them up or store them in baskets.

Looking down over the Herb Garden and Ornamental Vegetable Garden on 25 December 2010. The temperature outside was 18°C (-0.4°F).

Snow!

As I write this, high up in the hop kiln overlooking the box balls and Ornamental Vegetable Garden, the snow is swirling around the windows and fragmenting the garden below into a thousand soft white shards and spangles.

Snow is lovely but can cause great damage. It can break branches – especially of evergreens – and even topple large trees on a slope as well as crumple seemingly robust structures. Some years ago I put up a fruit cage, bought at great expense, and left the very light netting on after the last of the fruit had been cleared in autumn. An inch of snow that then froze onto the nylon mesh was enough to buckle and bend half the aluminium frame irreparably.

But snow does very little harm to most garden plants and in fact acts as an insulating blanket against icy winds. As it thaws it provides moisture, which makes a muddy mess for a few days but is an important source of water for the spring growth of many plants. In fact, many bulbs rely almost entirely on melting snow for their water supply when growing in their natural habitat. Although it is a myth that it is never too cold to snow, it is true that by and large it only snows in this country at around or just below freezing which is not a disastrously cold temperature for hardy plants. If the snow is more than a few inches thick it will then insulate the ground – and all the plants sheltering beneath its blanket – from any further drop in temperature.

Hardy plants can manage cold, often down to extreme temperatures such as -15°C (5°F), and can sustain cold for weeks or months of about -5°C (23°F). Half-hardy plants such as penstemons, many camellias, or salvias do not, as a rule, tolerate any temperatures below -5°C (23°F) but can withstand the odd touch of frost, and tender plants such as basil or zinnias will not survive below 5°C (41°F).

This temperature, 5°C (41°F), when averaged out across a full 24 hours, is the point at which most plants start to grow. They will grow with increasing vigour and speed as the temperature rises – providing they have sufficient water – until around 25°C (77°F), at which point growth starts to decline at about the same rate that it rose. Of course as the temperature rises and the rate of growth increases along with the size of the plant, the demand for water increases greatly – even though, away from the Tropics, heat is usually associated with lack of water. This is where the effect of snow stretches into summer as plants draw upon the reserves of melt water in the soil.

Winter cold is healthy only because plants prepare for it. Our long autumn and spring wean plants into dormancy and then growth. This is why sudden frost in spring or autumn can have such disastrous effects – especially in spring. A plant that might withstand a month of bitter sub-zero temperatures can have half its growth killed by a few degrees of sudden frost in May. The new growth is not expecting it and although the plant might be conditioned to withstand cold, the new shoots have not had time to put this conditioning into practice. Timing is everything. Late summer growth will not have time to harden before early winter frosts. Grow too early and the same thing happens again – but more so in the spring frosts of April and even May.

Most temperate garden plants have adapted effective means to counter cold. Deciduous trees and shrubs drop their leaves and stop all but minimal root growth. Herbaceous plants will survive frozen ground perfectly happily because they have shut down all growth and gone into a state of hibernation. Annuals die as plants but leave a mass of seed that will survive the cold and grow in spring.

But the greatest damage done by cold to a garden is from the wind. Plants suffer from the wind chill factor just as much as humans, and wind will also dry plants in cold weather as effectively as a hair dryer. If this is combined with frozen soil it will quickly kill the plant. Thus, a still night with a frost of -5°C (23°F) will do little or no harm to any hardy plant. But add a wind of just 15mph to this – scarcely more than a breeze – and the air temperature drops to -12°C (10°F) and quite a few plants will be damaged if not killed. Unless your garden is stocked only by plants adapted to life on the prairie or the steppes, every garden should have hedges and windbreaks that will baffle the wind and break it up, protecting plants growing in their shelter.

Winter pruning fruit trees

This is the perfect time of year for pruning. Short of torrential rain, which is misery beyond the call of duty, you can prune equally well in frost, snow or balmy winter sunshine. In fact, it is a job that I try and do on frosty days when it is impossible to work with the frozen soil.

The purpose of winter pruning is to clear damaged or overcrowded growth, letting more light and air into the plant and to stimulate renewed and more vigorous growth as a result. Every tree and branch will develop a 'leader' which grows longer and more vigorously than the rest of the plant. It does this partly by hormones that stimulate more vigorous growth but also by suppressing the growth of the side shoots below them. If you remove that leader the lower shoots cease to be suppressed and will grow more vigorously – until one of them becomes a leader and the cycle is continued. The more you remove the leader, the more the plant will bush out and thicken up.

This means that if you wish to curtail growth, leave the pruning to midsummer when the foliage is fully grown and before the roots start to store food for winter. So, all trained fruit such as espaliers, cordons and fans are pruned now to encourage new growth and replace any weak shoots and then again in July to restrain and reduce excess growth.

All this applies to any deciduous shrub or tree, whether fruiting or not, but do not prune plums, apricots, peaches or cherries in winter. These should be pruned in late spring or early summer and only if absolutely necessary.

Apples and pears follow identical pruning regimes. Most apples and pears produce their fruit on spurs. These spurs take two to three years to produce fruit. So if you prune them off every year you will have lots of whippy stems and no fruit at all! The idea is to establish a framework of branches with plenty of spurs. However, you need light and air to get to them so remove all crowded or crossing branches and if it is a small tree, cut out the leader so it is open like a goblet. It is usually better to cut a few branches right back than to snip away at all of them.

If you are growing standard trees – that is large apples or pears with a clear trunk of at least 2m (6ft) – be brave and remove the lower branches flush with the trunk as the tree grows. This can sometimes involve removing half the growing structure but will speed up the eventual formation of a handsome, balanced, standard fruit tree.

If you are training your fruit trees as cordons, fans or espaliers the majority of your pruning will be in the summer. Only prune in winter to encourage growth. Obviously in the early stages of training fruit there will be quite a lot of winter pruning but as plants get more established this will become less and less. When pruning to train growth there are two things to remember:

1. Resist the temptation to train a healthy stem that is in the 'wrong' place. Prune it away and encourage a bud in the 'right' place – ie right by a horizontal wire if you growing espaliers – to replace it.

2. Winter pruning stimulates growth so prune the weakest growth hardest. This is completely counter-intuitive, but always works.

FEBRUARY

Until the end of January I always wake to an absolute, dark silence but as we go into February there is the faint but distinct chattering of birds just before dawn which itself marches in earlier and earlier. I know that the majority of British people find February the hardest month to bear but I love it. Regardless of the weather or the state of the garden, spring is coming and the days that hang so heavy in the weeks up to Christmas are getting lighter in weight and duration.

The snowdrops are at their very best around the middle of the month – although this can vary by as much as a fortnight according to the weather, and the primroses in the Coppice respond enthusiastically to any glimmers of sun.

Hellebores are the grandest plants of February and you cannot fail to be charmed by them, especially *Helleborus* × *hybridus* in all its forms, modestly holding their astonishing faces to the ground. My hellebores have bred indiscriminately which does result in rather a lot of muddy, pinky-brown flowers, but I encourage this. Too much good taste is bad for you.

As well as relishing the flowers that are increasing by the day, February is a busy month. The days may be lengthening but they are all too short and there is much to do. The garden has to be prepared like a ship setting out for a long voyage. Turn the compost heap. Lay that path. Check the mower and the garden furniture.

In the vegetable garden I dig and, if the soil is dry enough, sow broad beans and plant onion sets. However I do not worry about this – the readiness of the soil is much more significant than the date on the calendar.

The potting shed and greenhouse become the centre of activity, sowing seeds, taking dahlias out of hibernation, chitting potatoes. There is a temptation to sow too much, and almost everything that will eventually be planted outside is better left until March, but I have an irresistible impulse to sow as much as I can. I try to finish pruning the pleached limes, espaliered pears and any top fruit in the Orchard, as well as the roses, late-flowering clematis and buddlejas. But none of this can be forced. Ice and snow are likely and flooding not uncommon. But some years it is so mild that we give the grass its first trim before the month is out and the intoxicating smell of cut grass as the sun sets after 6pm on a late February afternoon is one of the most exciting moments in the entire gardening year.

CROCUS TOMMASINIANUS

It is a reasonable rule of thumb to say that wherever grass grows well crocuses will also be happy. But *Crocus tommasinianus* originates from European woodland and is an exception to this kind of meadow planting. This makes it ideal for planting in a mixed border or in the shade of deciduous shrubs and trees. One of the problems of having bulbs in a border is that they get dug up whenever you plant around them. But if a handful of corms get dug up as and when you plant something else it is of no consequence. Just bung them back in the ground. Like the Lenten rose, *C. tommasinianus* does have a tendency to hybridise and make muddy or weak colours but at this time of year I would rather have a gorgeous sea of inferior flowers than a handful of exact and superior blooms.

About 10 years ago, in September, we planted thousands of the tiny corms in the grass under a group of field maples that make the extension of the Coppice. This is an oddly complete space yet has never found a name or label to identify it. I guess every largish garden has bits like this that escape classification – probably because they do not need it. Although names can be a bit pretentious and solemn they do serve a real purpose. A distinct part of a garden without a name is curiously incomplete and impersonal. To know something fully you need to know its name. Yet occasionally places escape categorisation and naming. They just are and a label is redundant.

So in this nameless square patch – a good, peaceful patch of garden – we planted 1,000 *C. tommasinianus* and another 1,000 'Barr's Purple' which is a *tommasinianus* cultivar. If these quantities seem enormous bear in mind that a crocus corm is marble-sized and you need tens of thousands to make a really dramatic effect. Planting this number is a bit of a slog, but I go for the turf-lifting method, lifting a square of turf with a spade, spreading a score of corms at random on the surface of the exposed soil, and putting the turf back over the top of them.

HELLEBORES

About 20 years ago I went into the local health food store in Leominster for some brown rice and came out with a carload of oriental hybrid hellebores, which became the nucleus of the Spring Garden. Since then I have gathered a few more species hellebores but the heart lies in the Lenten roses, *Helleborus* × *hybridus* (formerly known as *H. orientalis*), and I have grown to love their appearance that spans the icy clutch of winter right through to the end of March.

Hellebores are usually expensive to buy, especially named hybrids, but they are good value because they last for a very long time, will grow in almost any conditions and need very little care – and have some of the most spectacular flowers that any garden can grow at a time of year when flowers are thin on the ground. They also spread themselves very prolifically from seed, so that now my Spring Garden has hundreds of the oriental hybrids. The only downside to this is their tendency to hybridise, making muddy colours and rather more quantity than quality, but none are bad and every now and then a real beauty crops up.

Lenten roses have flowers that vary from the palest creams to the darkest purples, via bright green. When you buy an unnamed species *Helleborus* x *hybridus* there is no saying what the colour will be and there are people who have devoted lifetimes to breeding hybrids whose flowers are predictable. *H.* x *hybridus* will cross not only with itself in all its various hybridisations but also with a number of other species, so garden seedlings will always be uncontrollable, which is good if you like surprises. Dark flowers are trendy and certainly beautiful and as a rule seedlings showing dark staining on the stem will produce the darkest flowers. However I like the full range that the Lenten rose can produce from a rich plum purple to a delicate greenish ivory shade, speckled with pink smudges and dots.

The best way to enjoy the details of these flower variations is not on the plant itself because the flowers hang down (to make it easier for pollinating bumblebees to get at their pollen) but to carefully cut a selection on short stems and then float them in a bowl filled with water – like tabletop waterlilies. They last a surprisingly long time like this and make the most beautiful centrepiece for any table.

One of the features of the oriental hybrids is that although their colorations and patterns – as pretty as a speckled bird's egg – are seemingly random, in fact the plants retain them throughout their long lives, so it is quite possible to try and combine two and start a breeding programme. These colours are not carried on the petals but on sepals that form a protective casing for the bud and the flower. The

flower buds are formed in summer, some six months before they appear, although it is common to see some flowers in late summer, the buds jumping the gun at the expense of a display when wanted. This is especially likely to happen with young or recently divided plants.

Many of the oriental hybrids are clones, which means that they do not produce fertile seed and can only be propagated by division. But they do not like being disturbed and if you are not careful they never recover. Set against this risk is that, unlike propagation from seed, each section of the plant will grow into an exact replica of its parent.

Given that hellebores do not like being moved, it pays to get the planting right first time and choose a spot where they will thrive. Hellebores are broadly woodland plants and the essence of all woodland flowers is that they flourish in a situation that has some degree of shade, shelter, rich soil (from all those years of fallen leaves) and often flower early before the canopy of leaves fills out above them blocking all light. Many will tolerate quite dry conditions, especially in summer, after flowering. That might sound a complicated, demanding mix, but in fact most gardens are a type of mini woodland by default and are ideal for woodland flowers like hellebores. The one thing that all hellebores hate is bad drainage, but soil that is rich enough will drain enough and the addition of grit to even really heavy clay, as well as lots of organic material, will do the trick.

When I plant a good-sized hellebore I will dig a deep hole (they have long roots) and add a generous amount of garden compost or mushroom compost. This extra material gives them a healthy start and they respond well. Remember that they live a long time, so it is worth taking some trouble when you plant them.

Oriental hellebores are the stars of the February garden. They hybridise indiscriminately and although the results can be a rather muddy pink they are always charming and sometimes stunning.

Hellebore leaves, of whatever variety, are leathery and evergreen and stay on the plant until replaced by new ones in spring as the flowers fade and set seed. However, as they grow older they can be prone to a fungal infection called *Coniothyrium hellebori*, which creates a chocolate blotching on the leaves, which then turn yellow and die. They also tend to form a tangle above emerging flowers so it is best to remove all the older leaves as the new growth appears – which is a job I do as soon as convenient in the New Year. This will reduce a confidently bulky plant to a fragile, naked thing, but do not worry – the flowers will be all the better for it and the extra light and air will reduce the risk of black spot. Hellebore leaves are very slow to break down on the compost heap but it is better not to compost any infected leaves, burning them instead.

WINTER HONEYSUCKLE

We have a woody, sprawling and somewhat scrawny bush at the end of the Spring Garden, near the back door, which I treasure at this time of year. It is the winter-flowering honeysuckle and it has the best of all fragrances of any winter plant. A single sprig will fill a room with its delicate but haunting scent.

Lonicera fragrantissima is the best-known and most common of the winter honeysuckles and has tiny ivory flowers on its bare, woody stems (although in mild areas it will be almost evergreen) that would scarcely be noticed in the glory of a May garden but which earn pride of place in stark midwinter. It will grow perfectly happily in dry shade and does not need feeding or rich soil as this will only encourage a mass of foliage at the expense of flowers. *L. standishii* comes from China and is very similar but more compact and completely deciduous. *L.* x *purpusii* is the offspring of a crossing of these two parents and is generally reckoned to have a hybrid robustness that combines both their qualities, being very free-flowering and vigorous.

Lonicera fragrantissima, *the winter-flowering honeysuckle, is an ungainly, scrubby shrub but has beautiful, delicate flowers with a deliciously subtle fragrance.*

RHUBARB

I have a bed devoted to rhubarb in the Ornamental Vegetable Garden and the fresh rosy shoots are growing strongly now as the days start to lengthen. These were all planted in 1993 after I brought back a bag of discarded roots from a visit to a rhubarb farm near Wakefield in Yorkshire and are a variety called 'Timperley Early'. At this stage of the season I force a few plants under terracotta rhubarb forcers – which is a fancy way of describing anything that will exclude light whilst allowing room for the young shoots to grow, which, when grown in darkness, are sweeter and have much smaller leaves.

The shoots can be harvested from February to midsummer, when they should then be left to develop good foliage that will feed the roots to ensure a vigorous crop next year. And a vigorous crop is needed because it is glorious stuff however you serve it, be it as fool, crumble, pie, jam, wine or just plain stewed, preferably for breakfast.

All rhubarb is good but early rhubarb is best. Hence the proliferation of forcers in Victorian times, drawing forward the sweet, light-deprived first shoots by as much as a month. Rhubarb is generally not sweet at all and shares spinach and chard's metallic tang, which comes from a high level of oxalic acid. In fact, the acidic content is so high that it will kill dogs and leave humans with a dramatically active tummy if eaten in sufficient quantity, although only the leaves have levels high enough to be properly dangerous.

It is dead easy to grow. Being a member of the *Rheum* family it prefers rich, deep soil so enrich the soil with whatever goodness you have and let the young plants grow for a couple of seasons before harvest, so that the roots have a chance to develop. Pull sparingly for the first few years and always stop by midsummer. No amount of cold weather will harm them and a period of cold is necessary to trigger new growth – but damp can rot the crowns so mulch them thickly with manure or compost at the end of autumn but be careful not to cover them. They will be good for 10 years or so when it is a good idea to dig them up, divide the roots with a spade and replant the segments to stimulate more vigorous growth.

My 'Timperley Early' rhubarb has grown and produced delicious stems in the same corner of the Ornamental Vegetable Garden for 20 years.

BROCCOLI

We are now entering the season of one of my favourite vegetables. I love the way that it becomes top-heavily laden with leaf with so much plant for such a delicate harvest. I even like its slowness, steadily managing all that growth from sowing in March to its first tentative picking – in my unforced garden at any rate – in February. To invest a year of growth and weeding, staking, protection from slugs, cabbage white butterflies and pigeons has an unfashionable layer of trust built into the relationship between gardener and plant. Growing broccoli takes some imagination and trust to surrender a slab of garden for a harvest tucked three seasons ahead.

But it is worth it. Let me get one thing straight. The great bulbous green affairs that you buy in supermarkets or are ubiquitously served along with stolidly unchanging 'seasonal veg' are not purple sprouting broccoli but calabrese. Calabrese (simply meaning 'from Calabria') is a mini green cauliflower really, not bad to eat and matures quicker over a longer season. However, it is synonymous with healthy but dull eating, whereas broccoli is subtle, delicate and restricted to a season that starts in the New Year and ends at the beginning of May – although in my own garden I am lucky to make a picking in February and by the end of April it is bolting faster than I can pick it.

'Romanesco' has a lime-green colour to the florets and is very delicious and 'Broccoletto' is a fast-maturing variety with a single edible floret. But I am happy with common or garden purple sprouting. I want a big, glaucous plant on a stalk as thick as my forearm (and best staked early otherwise they sprawl and lean and crowd each other like a drunken choir) producing a mass of delicate little branching buds, rich purple, each no bigger than a marble (and some much smaller than that) but each packed with a sweet – almost tender – cabbagey tang that demands ritual celebration.

It certainly demands lots of space but I always underplant it with lettuce for the first three months or so of its growing year. To this end it is best to plant the seedlings out in a grid with each plant at least a metre apart in any direction. There is always a temptation to sneak them closer than this but resist it. You gain no extra broccoli spears for your parsimony with space and you lose the chance to grow a good crop of

Purple sprouting broccoli is one of the true luxuries of the vegetable garden. Its season is short and growing time long, but worth all the time and space it demands.

lettuce, radish, rocket or even spinach before it gets crowded out with those lovely crinkled blue-green leaves.

I sow mine just when the harvest of last year's sowing is hitting its stride, in mid-March. It is easiest to sow them in plugs, thinning to one seed per plug, and then potting them on into generous pots – at least 8cm (3in) – so they can develop a decent root system before planting them out in early June. I always add a good inch or two of compost to a brassica bed, working it in lightly, so the plants grow strong and lusty. Firm the young seedlings into the ground fiercely well – again, treat them like trees rather than cabbages. But a year later, treat their harvest like a luxurious treat rather than worthy but dull health food.

Digging

If you grow vegetables, the chances are that your soil has remained uncultivated all winter. By digging it now you will let air into it and, where appropriate, add organic material ready for breaking down and sowing and planting in a month or two's time. The important thing is to break up the compaction and let in plenty of air. Use a spade, not a fork, as this will take out larger clods and let in more air. A fork can be used to break up the soil nearer to the final cultivation before sowing. Do not worry about digging in annual weeds – they will break down in the soil and enrich it – but remove any perennial weed roots such as bindweed, ground elder and couch grass as you go.

The Spring Garden at the brief point in late winter when both snowdrops and hellebores are in flower together beneath bare trees and shrubs.

How to dig

1. Dig a trench one spade deep (a spit) and a spade wide, cleaning out the 'crumbs' or loose soil, but trying not to mix topsoil and subsoil. Put the excavated soil into a barrow. Move back the width of a spade and dig the next line of the bed, throwing the soil from it to fill the first trench. It will fall into it at an angle, sloping away from you at 45 degrees.

2. If you have any manure or compost, spread a layer over the soil in front of you that you have just turned over. Do not bury it in the bottom of the trench as it should remain in the topsoil where the feeding roots can benefit most from it.

3. When you reach the end of the area or bed, use the soil in the barrow to backfill the last trench.

Inevitably, the soil level will have been raised above the surrounding area, but do not be alarmed by this, as it will gently subside whilst retaining its new light structure. The weather over the coming month or two will help break it down further and the extra air will promote more bacterial activity whilst the lack of compaction will greatly improve future root development and growth.

Spring pruning

A number of plants are best pruned between the middle of February and the middle of March. These tend to be those that flower on new growth for which pruning hard will stimulate lots of healthy new growth and hence lots of flowers. But inevitably there are exceptions. Here are the most common plants for spring pruning and how to deal with them.

Clematis

When pruning clematis there is one really important consideration: when does it flower? The old rhyme, 'if it flowers before June do not prune' will get you out of most trouble, but clematis can be subdivided into three flowering groups.

Group one is early-flowering clematis (which means up to late May) and includes *Clematis montana, C. alpina, C. armandii* and *C. macropetala*. These tend to have many small flowers that are produced on growth made the previous summer. So if you prune in early spring you will not harm the plant but will radically reduce the quantity of flowers. Ideally trim as necessary (to shape and size) in June.

Group two is mid-season clematis (from late May to early July) and these tend to have much less vigorous growth and much larger flowers. They include 'Niobe', 'Barbara Jackman', 'Nelly Moser', 'The President' and 'H.F. Young'. These often flower twice, first on growth produced the previous year and again on new growth. The second flush is always of smaller flowers. If you prune hard at this time of year you will not have any early, large flowers but plenty in late summer. The best bet is to remove all weak or straggly stems now as well as all growth above the top pair of healthy buds.

Group three is late-flowering clematis (after mid-June) and includes *C.* 'Jackmanii' and *C. viticella*. All are multi-stemmed and flower on growth made in spring, so the previous year's growth should be cleared away in late winter. I always cut down to about 60cm (2ft) from the ground, leaving at least two healthy pairs of buds.

Always prune all clematis very hard – to about 15cm (6in) – when you first plant them. This will encourage healthy growth from the base of the plant.

Roses

There is a lot of mystique about rose pruning, but the reality is that they are all tough shrubs that can take a mauling by anything from secateurs to a flail cutter and bounce back. However there are three considerations to bear in mind when pruning roses.

1. **Hybrid teas,** floribundas and hybrid perpetuals: these flower on the current season's wood so they should be pruned hard each spring, removing all weak, damaged or crossing stems first and then pruning the remaining stems to form an open bowl of stubby branches. Don't worry too much about making outward-sloping cuts, but do always cut just above a bud. Remember to cut the weakest growth hardest.

2. **Shrub roses:** these need little pruning, just a once-over with a hedge trimmer has proven effective. I prune mine in early spring by removing exceptionally long growth, damaged or crossing branches and then leave alone. There is a strong case for doing this in late summer or early autumn.

3. **Climbing roses:** these can be subdivided into two groups, climbers and ramblers: a) Climbers tend to have single, large flowers covering the period from early summer right into autumn, and include 'New Dawn', 'Albertine' and 'Dorothy Perkins'. These are best pruned in autumn (although it can be done at any time till March), trying to maintain a framework of long stems trained laterally, with side branches cut back to 2.5–5cm (1–2in) breaking from them. These short side branches will carry the flowers on new growth produced in spring. Ideally a third of the plant is removed each year – the oldest, woodiest stems – so that it is constantly renewing itself. b) Ramblers have clusters of smaller flowers that flower just once in midsummer. These include 'Bobbie James', 'Rambling Rector' and 'Paul's Himalayan Musk'. These need little pruning but should be trained and trimmed immediately after flowering as the flowers are carried mostly on stems grown in late summer.

Buddleja

Buddleja davidii is by far the most common butterfly bush and can be pruned right back in late winter to a couple of healthy buds. This will mean removing 90 per cent of all growth to a knuckly framework that will support the new shoots.

B. alternifolia and *B. globosa* both flower on wood made the previous year, so – like early-flowering clematis – should be pruned back by a third immediately after flowering. Older stems can also be removed at this point right to the ground to encourage replacement growth.

Always use really sharp tools for pruning. Not only does it make life much easier but it also makes for cleaner cuts and therefore causes less damage to the plant. Sharp tools are also much safer. But treat secateurs with respect – I have seen some very nasty injuries as a result of careless use (including cutting the top of one of my own fingers off!). In practice, this means sharpening your secateurs as often as you would sharpen a kitchen knife. It also means buying secateurs that will hold an edge. Invariably you get what you pay for. Personally I hate anvil types of secateurs and use the by-pass type, which seem to me to make a much cleaner cut.

Loppers are also important. Go for a good quality pair but never strain them – if it cannot be cut easily it should not be cut at all. Use a saw instead. Nowadays there are superb Japanese pruning saws on the market ranging from small folding ones to really large ones and some on extendable poles. They all work with a pulling action and are very easy to control precisely. I find them invaluable.

A great deal of research has been done on the business of painting wounds left by pruning. The current authoritative view is that nothing you might paint on a wound does any good at all. By far the best course is to leave a clean cut and let it heal itself.

Until new foliage grows, pruning removes the plant of potential food so mulch all climbers and shrubs with a generous layer of compost when you have finished pruning. Clematis and roses, in particular, respond well to a really heavy annual mulch of compost.

Clear all prunings, and if you own or can hire a shredder, shred them before adding to the compost heap. Failing that, stack them in a hidden corner to provide perfect cover for beneficial insects and mammals.

MARCH

As I get older, I love March more and more. There is a real sense of a long and arduous journey nearing its end and the garden comes home to the gardener. No other month trembles with such promise and although the month can bite back with some of the coldest weather of the year, there is a real chance of some sunshine and a few precious hours outside working in shirtsleeves.

The strange thing is that it always takes me by surprise. Spring arrives. It is odd, alarming even, thrilling and utterly predictable. The really strange thing is that there is no actual marker for this. For weeks, even months in the sun-soaked south, we have had the more traditional signs of spring with daffodils, primroses, birds singing and nesting and even the odd blossom blooming, but this is not necessarily enough. These are outrunners for the arrival, sounds coming from over the horizon rather than the song itself. And then you go outside, sniff the air and you recognise it for what it is. Spring is here, and will not go away for months.

It feels like a prize for enduring the long winter months. For the first few days I bask in the springness, letting that little bit of extra light, the flowers peeking their heads above winter's parapet, the extra volume of birdsong and the smell – yes the actual, measurable scent – of plants growing, all filter into my system.

But then there is a sense of urgency. I need to get on. There is much to do and now is the time to get my act together and start doing it.

This is slightly disingenuous of course. I garden all year round when I can but there is a sense of housekeeping in this. Jobs are done because they need doing not because there is that irresistible, visceral urge to be outside and part of the growing emerging world, which is how I feel in spring.

An aspect of March that is undervalued is the tremendous evening chorus of the birds. As the sun starts to sink the songbirds whip themselves to a frenzy of territorial abandon and the March garden resonates to their lovely song.

What the gardener really wants in March is drought. A month of dry days, every one with a little more daylight, would be a horticultural treat. This is because the first focus of attention must be the soil and if the weather is dry then you can get at the soil to dig, plant, weed, prune – all the things that are impossible when it is wet. Dry soil also warms up much quicker than wet.

If your soil is ready then March is a good time to plant and move things around. But ready means, above all, warm enough. The only way to know this is by touch. Pick up a handful of earth. If it feels cold and clammy to the skin then seeds will not germinate and roots will not grow. If it feels warm, holds together when squeezed and yet can easily be crumbled, then it is ideal.

ANNUALS FROM SEED

I like bedding. It is cheerful and lifts our spirits after a long winter. It can be used in containers, hanging baskets, mixed borders or as a display on its own. Bedding is beautiful and fun.

Wherever you buy your plants, do not be seduced by a fine show of flowers, especially early in the growing season. More flowers in the garden centre means less in your garden. Look for strong, bushy plants and when you get home pinch off any flowers to rest the plant, allowing it to put energy into developing strong roots and foliage which in turn will encourage a better show in a few weeks that will last a good long time.

But growing your own from seed is much more economic, much more fun and will give you ten times the number of plants. For the cost of a tray of bedding you can buy up to half a dozen packets of seed, each one of which will give you half a dozen trays of plants. The range of seeds that you can buy is also far greater than the available plants.

Every year here at Longmeadow I grow tobacco plants (*Nicotiana sylvestris*), tithonia, leonotis, sunflowers, nasturtiums, marigolds, salvias, cleome, rudbeckia, cerinthe, poppies (Shirley, field, opium and the Californian poppy, eschscholzia), and zinnias. I also grow the biennials: wallflowers, foxgloves and antirrhinums and use them as bedding. Other gardeners will have an equally long list composed of entirely different plants.

When you buy bedding you are invariably presented with trays of mixed colours bred for their ease of production and ability to flower for a long time and resist disease. None of these qualities could ever be considered a bad thing, but it is geared first for the producer rather than the consumer. If you choose your own seed you can carefully select the colours, heights and fragrances that please you and work best in your particular garden. You are actively taking control rather than being on the receiving end of a marketing plan.

GROWING ANNUALS

There are three kinds of annuals: hardy, half-hardy and tender. These can be divided into those that will tolerate frost and those that will not. So for example busy Lizzies, petunias and nicotiana grow in response to heat rather than light and will not cope with frost, which means that at Longmeadow they cannot be planted outside until well into May (although in a city frost is much less likely after mid-April).

Hardy annuals like cornflowers, poppies and nigella can cope with cold – although none enjoy it. As a result it is often best to sow hardy annuals directly where they are to grow, knowing that the seedlings will survive a late, cold snap.

You can simply scatter seed into a border and let them germinate and grow where they fall. That is, after all, what nature does. But there is a risk that weeds will swamp them, that they will not grow precisely where you intended them to and also of being weeded out by you by mistake. It is easily done.

The time-honoured – and very effective – way of incorporating seed into a mixed border without it looking unnatural is to sow them in zigzags, crosses or circles – so you can see where they are growing and not weed them out – and then thin the seedlings so that the artificiality of these shapes is lost. It works every time. The only hard bit is to sow much more thinly than seems sensible and then thin the seedlings ruthlessly so that each plant has room to enrich itself – as much as 15cm (6in) for most plants. If the seeds are very small try mixing them with sharp sand or vermiculite to thin their spread.

Half-hardy or tender annuals must be sown under cover. Ideally the temperature will be fairly constant and stay above 6°C (43°F). In practice, this merely needs to be frost-free, and the harder they are grown – or the sooner they are exposed to the outdoor climate and weather – the healthier they will be. If you have a greenhouse then you have your own plant factory. But a porch or windowsill will do and a coldframe is a brilliant investment. If you are growing seeds on a windowsill remember that the seedlings invariably crane towards the light, so turn them daily. Also avoid a southern window as this can easily become too hot at midday.

I sow small seeds like tobacco plants in seed trays and use plugs for larger ones like sunflowers, but small pots sprinkled with seed are effective in a limited space.

Whatever containers you choose use a general-purpose peat-free compost and I recommend adding some vermiculite or horticultural grit to lighten it and improve drainage and root-run. Fill the container and gently level it off 1cm (½in) from the top. Scatter the seeds thinly over the surface. Lightly sprinkle a thin layer of sieved compost or vermiculite over the top. Then soak the container in a sink for a few minutes or water gently, which works just as well.

Most seedlings will emerge within two weeks. The growth of two true leaves is an indication that they have roots and can be pricked out into individual pots or plugs. Always do this holding them by a leaf – which is replaceable and tough – not the stem, which bruises very easily. Grow them on, being careful that they do not dry out but are not waterlogged. When they are between 8–15cm (3–6in) tall, start to harden them off in a sheltered spot outside, protecting them from cold nights, wind and very heavy rain. It is always worth leaving any bought plant – even from a nearby garden centre – for at least a week to get used to your garden's micro-conditions before planting it out.

Plant your seedlings out at the beginning of June making sure that they have plenty of light. Water them in well. Then leave them to get gloriously on with it.

DIVIDING HERBACEOUS PERENNIALS

Herbaceous perennials will grow much better if you divide them every three or four years, and at the same time you will produce a supply of free new plants. The best time to do it is just as they are starting into growth in spring.

If the plant has strong fleshy roots, like a hosta, the best method is to cut or chop them with a sharp knife or spade. Herbaceous plants have the strongest roots on the edge so cut them like a cake, ensuring each new slice is mostly outside roots. Make sure that there is a healthy bud visible in each slice. Discard the leftover central section as this will be exhausted.

If the roots are fibrous, like with helianthus or hemerocallis, then the time-honoured method is to prise the plant apart with two garden forks, back to back, levering one against the other – although I find it is often easier to tease the roots apart by hand. Again, discard the old central section and replant the more vigorous outside roots. Replant in groups of three to create clumps that will quickly grow together as one impressive display.

DIVIDING SNOWDROPS

The best way to speed up the spread of a clump of snowdrops and to encourage them to flower with renewed vigour is to lift them straight after flowering, divide up the mass of bulbs and replant them in smaller groups a few feet apart. Seed dispersal will mean that these clumps gradually meet, and repeated division every three or four years will quicken this process.

Carefully lift a clump with a fork, trying not to damage the roots too much, and tease out the bulbs individually or in small groups before replanting them. If planting in turf, leave an area of bare soil around each clump to help them establish.

I have planted snowdrops as 'dry' bulbs in autumn but without great success. It is much, much better to either buy them as plants or persuade someone you know to let you have a clump as they divide them. As long as they do not dry out too much until the foliage has died down they are almost certain to survive and flower and gradually spread.

DIVIDING AGAPANTHUS

Agapanthus do best with very good drainage and poor soil so I am preparing a mixture of potting compost with an equal volume of grit.

Agapanthus has fleshy roots that should be divided when – but not before – they completely fill the container they are in.

I have cut this plant in two with a sharp knife to make two healthy plants that will grow with renewed vigour.

After repotting – and in the case of agapanthus the container should be quite constricted even when freshly repotted – give them a good soak.

FRITILLARIES

Most spring bulbs do best in well-drained soil, which can be a problem for those of us who garden on heavy clay. However, one of my favourite spring bulbs positively relishes damp soil in winter and early spring. This is the common snake's head fritillary, *Fritillaria meleagris*. Its flowers have pointed bonnets hanging down – hence their folk name of sulky ladies – chequered with a patchwork of mauves, pinks, purples, greens and sometimes whites in various permutations, with some almost pure white and others richly purple.

I first planted a batch of 100 plants (more expensive than planting the bulbs but more likely to succeed) down the far end of the Spring Garden that is bounded on one side by water meadows. These meadows flood often and the water inevitably spills into my garden, sometimes dramatically, and the first place to get a soaking is that far spit of Spring Garden. It does not seem to bother the fritillaries at all.

I have also planted another batch up on the Cricket Pitch that will grow in the grass with crocus and narcissi. Although this does not flood and they had to endure an exceptionally dry first spring and summer, the ground is heavy enough to hold moisture and make them feel at home.

It is a native wildflower found naturally in wet meadows – the most famous of which is behind Magdalen College Oxford where thousands of them flower each spring. My section of flooded Spring Garden, therefore, is a home from home. The bulb goes dormant from June until August when it grows new shoots that stop just below the surface when the nights begin to cool. As soon as the weather warms up in spring they start to grow fast from this poised position so that they can flower and set seed before the grass gets growing. In this way they could – and very occasionally still do – co-exist with grazing cattle and haymaking on the same ground as long as they were not poisoned by fertiliser.

Snake's head fritillaries are the easiest of this wide-ranging member of the lily family, but there are more than 100 others to grow. All share one unlikely pollinator, which is the queen wasp, active as a solitary operator in spring. I have the crown imperial, *Fritillaria imperialis*, growing in the Spring Garden, Jewel Garden and Dry Garden and it performs heroically in each of these different spots.

Crown imperial fritillaries are from Kashmir and stand about 90cm (3ft) tall on curiously flattened, thick chocolate stems, the bright flowers hanging beneath a dreadlocked topknot of leaf. They start to fill the Spring Garden with their distinctive

The imperial fritillary is a majestic clown and the star of the Spring Garden in April. It needs good drainage and I always add plenty of grit to its planting hole.

aroma of fox and tomcat when the leaves first appear, halfway through March. As the flowers first open the air is rancid with fritillary fragrance, but then it seems to diminish. Rather like snake's head fritillaries, the words to describe them are touched with distaste but I adore them, smell and all, and think of them as one of the joys of my spring.

We have three types: the common one, which is the orange or brick-coloured version; the yellow *F. i.* 'Maxima Lutea'; and the deep orange *F. i.* 'Rubra'.

It is an ancient garden plant, appearing in Renaissance paintings and featuring in the very oldest British books on gardening. It was introduced in the late sixteenth century and is called the imperial fritillary because it was first grown in Europe in the imperial gardens of Vienna. Inside the petals are nectaries that were thought to be teardrops. A legend grew up in the Christian religion because it was believed that the imperial fritillary was the only flower that refused to bow its head on Good Friday and now hangs its flowers in shame. In Persia the tears were thought to be those of a queen whose fidelity was questioned by her jealous husband whereupon she was changed by an angel into the flower, the glistening teardrops forever expressing her regret and remorse.

One of its many oddities – and therefore attractions – is that you plant the bulbs (which are enormous) on their side. This is because they have a hollow central core through which the main shoot develops and if they are on their side then water cannot sit in the bulb and rot it. It is a good idea to lift the mature bulbs every few years and to replant them to induce renewed vigour.

The bulbs are good at making offsets and a few bulbs will gradually spread into a stand of tousle-haired imperial splendour.

PRIMROSES AND COWSLIPS

I am often asked what my favourite flower is. The truth is that the answer depends on the season, the situation, on what I am doing – and even who I am doing it with. There are lots of favourites. But whenever I am forced to stump up one floral treasure above all others, I always go for the common, humble primrose. No other plant so perfectly celebrates the coming of spring or does it with such gentle charm and beauty.

Primroses are woodland flowers, loving cool, damp banks and glades and thriving in the lee of a hedgerow or in coppiced woodland, in particular hazel. The native primrose is *Primula vulgaris* and although the name describes a very distinctive pale yellow, they occur quite naturally in every shade from almost pure white to a distinct orange. Primroses like wet soil best, with summer shade. The drier the local climate, the more they need shade and heavy soil that will hold moisture. So add plenty of organic material to help conserve moisture for the roots. Summer drought is not really a problem as long as they get plenty of moisture in autumn and the first part of the year. They will spread steadily, especially if exposed to light every few years, so if they are in a border make sure that they are underplanting shrubs that can be pruned hard every now and then.

Primroses are perennials that will last for a number of years but will spread quite fast by seed. I underplanted the hazels in the Coppice with a few dozen primroses (bought very cheaply as small plants) and these quickly spread to form great pools and clumps of yellow flowers nestling in their bouquet of lime-green leaves.

The date of this flowering is very variable, influenced both by weather and the gradual changing of our seasons. Some years they are at their very best at the end of February but after a cold winter I expect them to be perfect for the last couple of weeks of March.

Cowslips, *Primula veris*, are primulas whose coronets of small flowers are borne on single long stems. They are superficially similar to the primrose but very different in their preferred habitat. Cowslips are plants of open downland and meadow. However, they can easily be grown in a garden if you have a sunny patch of well-

A pure white snake's head fritillary, **Fritillaria meleagris**. *Most are a chequerboard of violets and purple. All are happiest in a a damp meadow or a damp, partly shaded border.*

drained grass that can be left uncut long enough for the flowers to set seed – which effectively means the beginning of July. I planted some on the sunny side of the Coppice and they have flourished, even though my soil is rather heavy. I bought mine as a tray of plugs, which is a very good way of buying wildflowers cheaply.

They have cross-pollinated with the primroses, which produces sterile hybrids known as false oxlips that have the tall stem of the cowslip with the larger flowers of the primrose.

All these primulas can be divided and moved and the best time to do that is immediately after flowering so they can be ready to regrow for as long as possible before flowering.

VIOLETS AND PANSIES

The sweet violet, *Viola odorata*, is one of the most quietly modest and yet beautiful of all flowers. Although perfectly suited to any garden and wonderful in pots on a windowsill or balcony, they are a woodland plant and by bringing them into the garden – or even to a pot outside the back door – you capture the essence of British woodland. We have them growing here in the Coppice and running under the hawthorn hedges.

Their innocent beauty is a by-product of a remorseless quest for pollination. In theory violets are self-fertilising, so any insect activity is a bonus. In practice, they depend heavily on insect pollination, especially by bees, and the early flowers rarely set seed although a later flush of autumn flowers set seed easily.

Pick a little bunch of violets and a room will be infused with their gentle and yet persuasive fragrance – a floral cologne. People have imitated this in toilet water, sweets and soaps but nothing can compare to picking a few flowers on their spindly stalks. The flowers rise on stalks as fine as those of cress seedlings from heart-shaped, fragrant green leaves that remain in a modest form all winter but are now creating new growth of a lovely freshness.

Although cowslips are native to chalky, open downland, which is as unlike the conditions at Longmeadow as could be imagined, they grow well in a dry strip along the sunny edge of the Coppice.

The leaves in turn sprout from a surprisingly knobbly stem, which throws out runners, just like strawberry plants, and new plants root along their length. This means that in the right conditions violets will spread quickly. They can be moved around easily, too. The best time to do it is immediately after flowering and a well-established clump should be broken up every few years to encourage it to spread further faster. You will also find that any group that is apparently healthy but not flowering can be spurred into blooming by lifting and moving.

Although wild and tough and needing almost no cultivation, violets are prone to attack by red spider mite, which will indicate itself by yellowing leaves. Poor drainage and heavy soil will exasperate this situation, as will too close planting.

I like to grow pansies in terracotta pots although they can be planted in any moist, rich soil. They hate dry conditions but should not be in a bog so, as ever, plenty of compost or leaf mould in the ground is ideal. It is best to plant pansies in autumn although I often leave it until February or March. Whenever they are planted you should pinch out all flower heads. This might seem harsh but it will give the plant a chance to establish roots before putting its energy into the flowers – and as a result you will have many more flowers for much longer. Regular dead-heading will do much to increase the length of display.

As spring progresses the plants will grow increasingly leggy and as the supply of flowers diminishes the entire plant should be cut back to one joint above soil level. However many of us treat pansies as annuals and consign them to the compost heap at this stage.

Pansies and violas can be raised from seed sown in June or July with the young plants placed in position in autumn, or from cuttings taken in late summer. The advantage of cuttings is that they preserve the exact particulars of the parent plant, whereas seed will always be an unknown combination of both parents.

As March progresses, the propagating greenhouse starts to fill with seedlings. Although they are watched over carefully, I resist overwatering or feeding as all must grow hardy enough to cope with life on the outside.

GETTING STARTED

At this time of year there is a temptation to rush out into the garden on the first sunny day and sow masses of seeds in the full expectation of a marvellous harvest by mid-spring. Stop.

Feel the soil. If it is cold to your touch then few seeds will germinate. Whilst there is no point in delaying things if the conditions are right, there is little to be gained by trying to cheat the seasons or the weather. One of the real skills of growing good vegetables at home is in paying attention to and working with the particular conditions on your plot.

As soon as your soil is dry enough to rake without the soil sticking to the tines and does not feel cold to touch, you can sow carrots, parsnips, broad beans, rocket and spinach. You can also plant onion and shallot sets, burying them so that the tops are clear of the ground. The only crop that needs getting into the ground as soon as possible is garlic – ideally it should be in the ground by Christmas – but it is worth a go, particularly if you live in a cold area.

A good rule of thumb is that if the weeds are not growing then it is too cold for your vegetable seeds. Prepare the soil and leave it. When a flush of weeds starts to grow, hoe them off and sow your seeds – in this way your seeds will avoid competition in their first vital weeks of growth. When you sow outside you can make drills and sow the seed in rows, which means you can see them as soon as they emerge and avoid treading on them or confusing them with weeds. Alternatively you can broadcast the seed. I use this latter technique for carrots and also when I use a seed mixture such as 'Saladesi', or 'Saladini', which I broadcast and do not thin except to eat. Otherwise I stick to rows.

However you sow, always be careful to sow as thinly as possible. In an ideal world you would have the seeds twice as close together as the final desired spacing. This would provide for failed germination and the thinning out of any ailing seedlings, along with normal thinning. In practice, when sowing direct, just keep it thin!

Thin them carefully as soon as they are large enough to handle, thin again a few weeks later (eating the meltingly tender thinnings, roots and all) so that you are left with a row of maturing plants about 8–23cm (3–9in) apart.

The advantages of this are that the roots of the plants are not disturbed more than they have to be and it requires no potting compost, seed trays, plugs, greenhouses or paraphernalia of any kind. As long as the growing medium is well drained and quite rich – I always add and lightly rake in 2.5cm (1in) of fresh garden compost before sowing or planting out salad crops – they should grow well. The disadvantage is that they depend on the right soil conditions and are very susceptible to snails and slugs, especially at the young seedling stage and especially at this time of year when growth can be slowed almost to a standstill by a bout of cold weather.

It is just as easy and much more controllable to sow the seed in plugs or seed trays, grow them into good-sized seedlings and then plant out at 23cm (9in) spacing when the conditions are right and they are big enough to withstand any kind of slug or snail attack. A proprietary peat-free general-purpose compost will be fine for this.

March is the perfect time to sow seeds under cover for planting outside about a month or so later, when the soil has warmed up a little and – crucially – there is more daylight. A greenhouse is best but coldframes are very good and a porch or spare windowsill or two perfectly workable. Even if you have a small yard and intend just to grow a few pots of salad leaves at any one time a small coldframe is a really good idea. You can make this yourself with a wooden frame lined with polystyrene insulation board and using polythene or Perspex for the cover. This will provide just enough extra protection to raise almost anything from seed.

There is a range of salad plants adapted to grow well in cool spring weather but which does not thrive so well in the heat of summer. This is what I concentrate on in early spring.

Salad crops for early spring sowing:

- rocket
- mizuna
- mibuna
- corn salad
- land cress
- lettuces 'All the Year Round', 'Tom Thumb', 'Little Gem' and 'Rouge d'Hiver'.

CHARD

No other vegetable gleams like chard. A healthy leaf is as lacquered and glossy as holly, and the stem buffed to the point it looks molten. Rhubarb chard has green leaves and red stalks; ruby chard has both red leaves and stalks – with the crimson intensity of cut beetroot; and rainbow chard has stems of yellow, orange, pink and red. Once cooked, all taste remarkably similar, so for reliability and taste I would opt for the green leaves and white stems of Swiss chard.

I sow the first batch about now and then a subsequent one in early August that will provide plants to overwinter for the following spring. The seeds can be sown direct but I sow them in plugs or blocks restricting one seed per unit, growing them on and hardening off before planting out at 23cm (9in) spacing. They like a really rich, moisture-retentive soil.

Being biennials they will only go to seed in the first growing season if they are distressed, so the consistency of water supply is as important as the quantity. If some do start to bolt then I cut the central stem down to the ground and give them a soak.

We shred the green leaves from the stalks and cook and use them exactly like spinach – although the leaves do not have such a strong after-taste – and the stems are very good cooked separately in water or stock and served either with oil and lemon or a béchamel sauce. The leaves and stalks chopped up together make a very good filling for a pie or flan.

CROP ROTATION

Rotating the crops around your plot, whatever its size, will help avoid the build up of pests and diseases, increase fertility and enable you to keep your soil in excellent condition.

Vegetables are usually divided into three groups that share the same cultivation needs. Within the demands of space and the desire to keep the garden looking attractive at all times, it is worth trying not to mix the elements from each group and to sequence them in the following order.

Group one: legumes

Legumes include all peas and beans – and the fruit vegetables are usefully added to this group which includes tomatoes, peppers, cucumbers, aubergines, squashes, pumpkins, sweetcorn, courgettes, okra and celery. I also grow lettuces alongside these crops as space is made available. Peas and beans have the capability of 'fixing' nitrogen from the air and leaving a residue of it in the soil. This means that plants succeeding legumes can tap into extra nitrogen, which will encourage green, leafy growth.

Group two: brassicas

Brassicas include all cabbages, cauliflowers, broccoli, Brussels sprouts and kales – all of which benefit from being grown in soil that has just been cleared of legumes. I add a thin layer of compost before planting, which I work lightly into the topsoil, to encourage extra growth. Swedes, turnips, radishes, kohl rabi, land cress and mizuna are also brassicas and should share the same rotation. Brassicas are subject to clubroot which thrives in soil with a low pH, so lime is often added before planting to soil that is inclined to be acidic.

Group three: roots

Nitrogen will encourage leafy growth at the expense of good roots so root crops usually follow on from brassicas, which will have taken up all the nitrogen left by legumes, and no extra compost is added to the soil. Although compost is not added, the soil should be worked deeply to ensure a free root run. This group includes carrots, parsnips, beetroots, onions, leeks, garlic, salsify and scorzonera, but not turnips and swedes, which are group two. In practice, this is the most diverse and confusing group. Carrots and parsnips, for example, grow best in soil that has had no added manure or compost incorporated whereas it is important to top up the soil with compost for celery, celeriac and potatoes.

If in doubt over when and where to enrich a plot as part of the rotation, a thin layer of good garden compost worked lightly into the topsoil will always do a lot of good to the soil and no harm to any crop.

LEGUMES

Whatever your soil is like, growing legumes of any kind will improve it for future crops. The leguminous vegetables that are easy to grow in British gardens are peas, broad beans, French beans – both dwarf and climbing – and runner beans. Leguminous green manures include lupins, clover, vetches and trefoil.

All legumes host bacteria – rhizobium – that can convert the nitrogen that makes up around 80 per cent of our atmosphere into a form that can be easily absorbed by plants. In return, the plants feed the bacteria sugars.

Nitrogen is essential for green, new growth in plants but too much creates an excess of soft, sappy growth, which in turn attracts aphids and fungal problems.

Hence a legume crop every two years (three growing seasons) maintains the right balance if they are followed by brassicas over the subsequent autumn, and winter and by root crops after that.

PRUNING CURRANTS AND GOOSEBERRIES

Now is the time to prune red and white currants as well as gooseberries (but not black currants, which should be pruned in August after harvesting the fruit). White currants are effectively albino versions of red currants and are grown under identical conditions, while gooseberries can be treated the same way. All three produce their fruit on knobbly spurs rather like apples. These spurs need a permanent, open framework of branches.

You can grow them in two ways, either as a bush, raised up on a short stem like a wine glass, or as a cordon. If growing bushes, begin by taking out all inward-growing branches and any that are damaged or crossing. The aim is to reduce the bush to an open bowl made out of four or five uncrowded ribs standing on a clean, straight or concave stem. Then prune these remaining branches by about a quarter to a third so you have a strong, stumpy framework with plenty of spurs. The idea is to create maximum ventilation and light, both to ripen the fruit and to deter sawfly that hate exposed, windy conditions.

Gooseberries can also be grown as cordons. A cordon is a single stem grown vertically either supported by wires or against a fence. It enables you to grow many varieties in a limited space. Each March you should cut off all lateral growth right back to the woody spurs that will carry the fruit, and thin these so that they grow as a series of knobbly protrusions with at least 2.5cm (1in) between them. It looks dramatic but will encourage heavy fruiting.

Gooseberries are tough plants and almost relish harsh treatment. But they are a subtly fragrant and delicious fruit.

The first light trim of the year. The mown grass at Longmeadow is not pampered and serves only to provide a smooth green surface. Any weeds that fit that bill are welcome.

Lawns

Although admittedly male, middle-aged and obsessed with gardening I cannot say that I share the need for a perfect lawn. If it is green, crisp and even then it will do for me. I really do not object to the presence of a few daisies, dandelions or some moss. If an area of grass is pleasant to sit or lie on then it is doing its job. But even I like to see grass neatly mown, whereas my wife claims that she would only mow paths wide enough to walk down and let the rest of the grass grow meadow-long. I suspect that this is a gender thing. Lawns bring out the martinet in men and meadows the romantic in women.

But most of us would agree that an expanse of green harmoniously links any of the colours that border it. Lawns and grass paths make the perfect balance between the business of borders, trees and hedges.

To get a 'good' lawn you have to think positively. Put your efforts into healthy grass rather than fighting perceived problems such as daisies, moss, ants, worm casts, moles, plantains, dandelions and fairy rings. Nine times out of ten if the grass is healthy then everything else will look after itself.

The best grass likes well-drained soil. Moss, for example, is always a symptom of poor drainage, made worse by shade. Unfortunately even the best-prepared soil becomes compacted by matted roots, rain and, especially, normal family use. The answer is to work on it at least once a year by sticking a fork in the ground and wiggling it about and repeating the process every 15cm (6in) or so. Then mix up equal portions of sieved topsoil, sharp sand and sieved leaf mould or compost. (If you

STIGA

do not have these things to hand then just sharp or silver sand will do the job.) Spread it across the area you have pricked and brush it in with a stiff broom, filling the holes with the mixture. This will help drainage and feed the grass.

It is also worth giving the lawn a good scratch with a wire rake. This will get at all the overwintering thatch and moss, and let light and water get to the soil and to the roots of the grass. Put the debris on the compost and then mow.

Level any dips and hollows by lifting with a fork (to get rid of compaction) and then filling to level with a 50:50 mixture of sharp sand and sieved compost. Sow some seed over this and try to keep animal and human feet off it: within a month or so it will be fully integrated.

When it comes to mowing the lawn the most harm that you can do is to cut it too short. The grass will be a lot healthier if allowed to grow to at least 1cm (½in) and preferably a bit longer. Do not take too much off in one go. A light trim that evens the grasses to the same length will make a dramatic difference – and be much quicker than a less frequent scalping. It is also important to collect the clippings except in very dry weather. They will do more good as part of the compost heap than left on the lawn – even when finely chopped.

Mulching

Mulching mixed borders is probably the single most useful job to be done in early spring. Not only will it feed the plants and improve the soil but also nothing is so useful in combating drought or suppressing weeds.

Before you mulch, cut back the last of winter growth, do any pruning that has not been done and move anything you want to reposition as well as removing every scrap of perennial weeds. A day spent preparing the border like this is never wasted and by the time you get around to actually spreading the mulch you are thoroughly reacquainted with the spring-cleaned face of your borders.

Mulch controls weeds by denying them light. Many annual weeds are triggered into germination by a combination of light and disturbance – which is why it is a good idea to leave a cleared piece of ground for a week or two before planting, to let the weed seeds germinate and then be hoed off. With annual weeds, like goosegrass, chickweed, and groundsel (which can grow and set seed within an amazing five weeks), emerging seedlings will not develop and the seeds will not germinate. Perennial weeds will grow through mulch but are much easier to pull up.

Whatever you use to mulch (other, of course, than an artificial mulch) must be at least 5cm (2in) thick to control weeds. Less than that is not worth doing. It is better to do half an area properly than the whole thing too thinly. Mulching to a thickness of 10cm (4in) is ideal. That is quite a lot but the more densely a bed is planted the less mulch you will use, and do not skimp on the thickness simply to make it go round.

It is astonishing how effectively mulch works into the soil, transforming its texture. If the soil is light it will give it body, helping it to hold water, and if it is heavy it lightens it, helping it to drain and the roots to grow more easily. It does not, of course, mix in all by itself. Earthworms in particular, but the teeming underworld of subterranean life in general, take it down, digest it and incorporate it into the humus.

The business of feeding plants is hugely over-estimated and is the least important aspect of any mulch. A fertiliser is only necessary when the soil is deficient in one of the three main nutrients: nitrogen, phosphorus or potassium. A balance of the three, regardless of the actual levels, will create healthy growth, albeit at different rates.

Obviously green growth, especially in spring, will lead to good photosynthesis and a healthy plant, so the low levels of nitrogen in compost or manure (about 2 per cent in poultry manure and 0.5 per cent in cow manure – with all other manures somewhere in between) will do some good. But the effect of garden compost opening out the soil and getting the water retention just right will do more for the plant – so that it can better tap into the soil's rich and complex sources of nutrition – than the direct effect of the fertiliser. Other than spraying any sick plant with a liquid seaweed solution I do not fertilise my outdoor plants at all.

Plants grown in a container will, however, outgrow the nutrients in the compost after a while and may need very weak supplements later in the year – but I find that liquid seaweed or comfrey is all I need for this.

I have experimented with different mulches over the years and any well-rotted organic material, from wood chips to cocoa shells, will do a good job, but I now just use homemade garden compost or bought-in mushroom compost. Garden compost is definitely the best mulch to use if possible because it is free, already on site, specific to your garden and is coarse enough to last for a year without degrading into the soil. It also goes some way to recycling nutrients that last year's plants have taken from the soil. But it is almost impossible to make enough compost for a year's supply of mulch in the flower borders, soil improver in the vegetable beds and for home-made potting compost, so I keep it primarily for edible plants and potting compost and buy in mushroom compost to mulch the flower borders. This is particularly good for

lightening heavy clay soil but is alkaline so if you have ericaceous beds or an already very alkaline soil you should not use it in successive years. Mushroom compost is a by-product of growing mushrooms so find a mushroom farm near you (preferably organic) and they should be more than willing to supply you with what is for them just waste.

Slugs and snails

I guess that there is not a gardener in the country who does not suffer from the predations of slugs or snails at some stage in the year. They certainly dominate all queries that I receive about garden pests. If there is an easy solution to this then I do not have the answer. But by managing the situation intelligently you can definitely reduce the extent of the problem.

There is an armoury of weapons to use against slugs and snails although none is entirely effective. Slug pellets are probably the worst. Do not use them. They are a poison and undeniably harmful to the garden's eco-balance and if you wish to garden organically they are not an option. Traps and barriers like beer, grit and eggshells help but only to a limited degree although I have used old copper piping as a barrier around seedlings outside and that certainly seemed to help.

The parasitic nematode *Phasmarhabditis hermaphrodita* is a minute worm that infects the slug or snail with a fatal disease and, rather gratifyingly, eats it up from the inside. You buy them live, mixed with a clay-based agent that must be diluted with water. The solution is then watered onto the specific area you wish to protect. The soil must be warm and wet and they need time to breed in order to work, remaining effective for around six weeks. Ducks, chickens, toads, hedgehogs and beetles eat slugs and thrushes love a snail. Longmeadow is full of all of these and we have practically no slug or snail problems at all – and take no preventative action. So all these should be encouraged in your garden. Rotovating bare soil in winter chops up some slugs and exposes others to hungry birds. All these things help.

Mulching with any organic material such as garden compost is the best way of suppressing weeds, retaining moisture and improving the soil's structure by adding organic material.

But nothing will stop slug and snail activity. For a start you are outnumbered. Go out into the garden on the next warm, damp evening and shine a torch around. The chances are that the garden will be slowly writhing with slimy bodies. In one experiment 27,500 slugs were taken from one small garden without making a noticeable difference to slug activity. Densities of 200 slugs per square metre are moderate. In light of this it is amazing how little damage they cause to our gardens.

The biggest slugs do not necessarily do the most damage. There are four main garden types: the grey field slug (*Deroceras reticulatum*) will eat anything and reproduces three new generations a year.

The garden slug (*Arion hortensis*) is shiny black with an orange belly and omnivorous. Its party tricks are to eat off bean plants at ground level and riddle potatoes with holes.

The keeled slug (*Milax budapestensis*) is probably the biggest problem although often very small. Also black, with a thin orange line down the centre of its back. It spends most of its life underground feeding off root crops but it will also eat what it can when it surfaces.

The black slug (*Arion ater*) can come in almost any colour but is differentiated from all others by its size, which can reach 20cm (8in) long. It is the least harmful of all garden slugs. Rotovate to get at the keeled slugs, and hand pick the field and garden slugs. Leave the poor black slug alone.

Snails love dry places to hide in and brick walls best of all, especially with crumbly or loose mortar. Apparently snails regulate their population density and growth rate by the amount of slime trails left by the adult snails. Where the ground is thoroughly slimed the young stay small, waiting for a window in the slime quota so that they can grow and slime about themselves. So by collecting hundreds of adult snails in one night you are going to cause an explosion of snail growth and reproduction.

But the starting point has to be one of reasonable co-existence. The first line of defence, whatever type of garden and wherever it might be, is to raise healthy plants. Plants that grow in a soil with good structure, nourished with garden compost made primarily from plants that have grown in that soil, that are not in any way forced against the weather, location or size, are in my experience remarkably trouble free.

The second factor, completely interconnected with healthy plants, is to encourage and sustain a balanced ecosystem in your garden. This is what inevitably happens in the natural world if man does not interfere. Leave your garden untouched and a balance will establish itself. But a garden is an ordered, unnatural place. One of the necessary horticultural skills is to assimilate that tight human control into a balance with the local wildlife. Make the garden a comfortable, easy place for your plants to grow rather than indulging in an ego trip about how you manage to grow such an unlikely range of plants in hostile conditions.

Slugs like young, soft plant tissue. So the trick is to make it available for as short a time as possible. Resist sowing or planting tender plants too soon. It is better to wait another week or so than risk losing plants, and do not feed plants any more than is absolutely necessary. Feed the soil not the plant. This will avoid a spurt of soft, sappy growth that slugs love. Grow your plants as 'hard' as possible, although do not stress the plant too much because plants that have been damaged or exposed to too much cold or drought are also the first to be attacked. Think of it like a soldier going to war – you want them to be fit, tough and able to withstand hardship – not ill and run down. They will be much tougher and better able to resist every kind of problem as a result, if not as big as conventional show plants might be.

Vegetables are the biggest problem because we inevitably grow them unnaturally, in neat rows and nurtured to look and taste as good as possible. To an extent this can be countered by mixing up your veg. Accept a level of slug activity: by midsummer I find at least half a dozen slugs in every lettuce I cut, but the lettuce is perfectly good and I just wash it carefully before eating it.

APRIL

There comes a time every year, somewhere between the middle of March and the first week of April, when I am possessed by my garden. It runs through my veins like a dancing river, occupies my sleep with dreams that sprout new leaves, and distracts me from any honest labour. I am bewitched by spring. And if I try and analyse it – which is not easy because this is an elemental feeling right inside my bones rather than any rational working of my mind – the emerging flowers are not the agent of this madness. No, the thing that seduces me in spring is the simple greenness of everything.

It is, of course, not just any green. I live in the country surrounded by green fields all year round, but when March slips into April I realise I have never truly seen green before. It works every time, catching me breathless and dazed by sheer exhilaration. This luminous lime green begins as a suggestion, speckling the hedgerows, almost shimmering above them rather than growing out of the branches, then starts licking up from the bare soil with shocking intensity. Certainly nothing so characterises the months of March, April and May as the range of greens shining like stained-glass windows.

As I get older I enjoy spring more and more. Without question it is my favourite time of year – but my enjoyment seems to start earlier and earlier. I used to think that May was the best month, but this shifted to April some years ago as it seemed to contain an extra edge of expectation – like the pleasures of Christmas Eve rather than the day itself. The pleasure is all in the process and the hope that is invested in the tentative opening out of the natural world. In fact, with climate change, the opening has become a lot less tentative than it was. April is now the greenest month and by its end, all the hedgerows are out and clothed in blossom.

And April is the busiest month for a gardener, especially after a winter like the one we have just had. So much to do and so little time to do it in. But every day is getting longer and the changing of the clocks at the end of March transforms any gardener's life. The extra hour at the end of the day is the best gift a gardener can have.

The longer evenings, that first genuine heat from the sun, the frenzy of birdsong and the way that the leaves seem to be visibly unfurling all focus energy into the garden this month, as well as the poor gardener who has bided the winter months. Nothing makes me happier than an April evening spent preparing the ground and sowing seeds for a summer harvest, while the garden settles gradually around me. Warm soil on my hands and the quiet rhythm of working at simple tasks that have been handed on from generation to generation is genuinely the best that life has to offer. In these early spring evenings the entire summer is poised and everything is possible.

There is a note of caution however. Whilst some Aprils, like 2011, can be warm and sunny, here at Longmeadow the soil is slow to warm up and a few sunny days are not enough to trigger things into life. I have gambled many times with the soil being warm enough to trigger healthy growth and so planted out seedlings that have been raised in the greenhouse and coldframe – only to see them sit sullenly in cold, damp soil, refusing to grow and eaten by slugs and snails. The only reliable indication that the soil is ready is the touch of your own hands. If it feels at all cold then it is not time – regardless of what the calendar tells you or what everyone else is doing. Trust your touch. Intuition counts far more than technical knowledge in the garden.

Narcissi growing in the Coppice above the new foliage of the primroses – and before the emerging leaves of the hazels shade out too much light.

The large perry pear tree in the Orchard is one of the first and best to come into blossom.

BLOSSOM

Blossom works on us in a way that few other flowers do. It is partly promise – that froth of flower is as life-affirming as anything your eyes will set on for another year; partly surprise – trees in flower are scarce in this country and it continues to amaze us that it happens at all; and partly the hint towards fecundity, because all blossom is merely a precursor to possible fruit. But, however it happens, we all know that heart-lurch of joy when we come across the first blossom of the year in the spring sunshine.

The *Rosaceae* family includes all our blossoming trees, so all roses are blossom and all blossom are sap-brothers under the bark. It explains a lot. If you see a cross-section of an apple, plum and dog rose flower they all look remarkably similar, with the fruit waiting to swell just behind the petals.

If you have a fruit tree that always has plenty of blossom but no fruit it is likely that you need a pollinator. As a rule it is best to grow three different types of apple or pear, a crab apple, a 'Morello' as well as a sweet cherry, more than one quince or medlar, a damson as well as plums, all of which will ensure pollination and therefore fruit.

Here in this garden and in the countryside of the Welsh Marches around us, the blossom unfolds as a procession, moving slowly through the landscape like the gorgeous retinue of a benign emperor. First is the blackthorn, *Prunus spinosa*, more often playing a minor role in a roadside or field hedge than planted in a garden. It is a plum of sorts and the tiny white flowers convert into sloes that wither the inside of your mouth with their astringency, although they make sloe gin (which is

the best winter-warming drink that there is). It has the wickedest thorns found in this country and, if trimmed, makes an impenetrable hedge. Just after the blackthorn comes the damson blossom, also pure white, dotted along the hedgerows between the cathedral cities of Hereford and Worcester like crenellations.

The plums follow just after damson, overlapping by a few days, just as the fruit is changed from strong and small to full and sweet by the process of breeding and hybridisation. Plum blossom sits on the trees thinly, measured as much by the sky between the flowers as the massing of the blooms themselves.

Next is the wild cherry, the gean, *Prunus avium*. I have a couple of these growing in the Coppice. They make large trees, up to 20m (65ft), often seen at the edge of a wood. *Prunus avium* 'Plena' is much heavier with flower than the gean and probably the one to plant in your garden. It also flowers a month later than the fully wild version, along with the sweet cherries. Ornamental cherries are the most popular of all flowering trees on account of their range of blossom, from the almond-like *Prunus × subhirtella* to the hanging white flower bells of 'Taihaku', which we have in the Damp Garden.

Crab apples and pears all come into flower at around the same time, according to variety. Certainly they always precede the earliest apple blossom. Quinces are next (not the much earlier chaenomeles but the tree, *Cydonia oblonga*). Quince blossom is one of my favourites, having flowers with an entrancing delicacy and sublime fragrance. Crab apples are also wonderful in a much more cheeky-chappy way. I don't know quite how they create that effect as they have perfect pink and white flowers that can fill the air with their scent, but compared to the bone-china refinement of quince or the purity of pear blossom, that is the result for me.

Pear is my favourite blossom, either in a great tree with the blossom piled up cumulus-like, or stretched out along the espaliers in the Ornamental Vegetable Garden. Like the damsons, bullaces and plums, pear blossom is always dainty. The flowers crowd into the branches but somehow politely so, not jostling but sharing the air. They are always pure white and carried whilst the leaves are only just beginning to open. In fact, the first flowers arrive on completely bare branches and the last are frilled with the green of the emerging leaves. Around this part of the world there are still a few – but rapidly diminishing – perry orchards with huge pear trees, hundreds of years old, as big as beeches and when they are in flower on a blue-skied day it is a marvel to rank with the best of the large magnolias, tulip trees or any flowering, growing thing on this earth.

The first apples in my orchard, like 'Tydeman's Early Worcester', which has magnolia-pink flowers, start in mid-April and the last, 'Norfolk Beefing' growing by the compost heaps, finishes fully a month later. The real difference between pear and apple blossom, other than the pure whiteness of pear and the 'strawberries-and-cream pink' of most apples, is that apple blossom is borne as the leaves start to emerge. This gives apple a fullness and softness that epitomises the fecundity of blossom-time. Also by the time that the last apple blossom is coming out the grass is growing strongly, the buttercups are out and the swallows are turning around the sky.

COPPICE FLOWERS

Towards the end of last century I planted an area of hazel coppice in my own garden, two blocks of perhaps 9 × 9m (30 × 30ft) each divided by a narrow path, and planted with hazel, ash and cherry standards. Beneath these trees I planted a range of woodland flowers but it quickly became apparent that some were more at home than others – and I have allowed these to dominate.

To coppice a shrub or tree means regularly cutting it right to the ground. Almost all deciduous trees as well as the native evergreens of holly, yew and box will take this treatment and regrow perfectly healthily for hundreds and even thousands of years without any ill effect. Indeed coppicing a tree is a way of making it live up to twice as long. This provides a harvest of wood for beansticks, thatching spars, fencing posts, firewood, charcoal and, historically at least, a hundred other uses, as well as stimulating the plant to regrow vigorously. The standards are trees allowed to grow tall for timber for building or furniture. You do not want too many of these lest they shade out the regrowth of the understory. In fact the effective density is much the same as a garden with one or two large trees that provide welcome shade without blotting out the light.

The secret is regular dramatic pruning. It looks brutal but is the key to survival for all these plants and the animal life that lives amongst them. By cutting the understory to the ground at regular intervals (I cut my hazel every seven years) you not only let in a great flood of light to the ground but also create a sophisticated and subtle evolving habitat with diminishing light and increasing growth between each cut. Many birds, insects and plants have evolved specifically to make the most of this, including butterflies, nightingales and my favourite woodland plants such as violets, primroses, bluebells, wild garlic and wood anemones.

The first coppice flowers to appear are primroses. By primrose I mean *Primula vulgaris*, with its subtle and delicate shadings of lemon, which grows best in hazel coppices and the mossy banks of lanes and tracks across this country in early spring. These are my favourite flowers of all, with more charm and innocence in a little clump beneath the shade of a bush than a massed border in full summer swing. They like wet soil best, with summer shade. The drier the local climate, the more they need shade and heavy soil that will hold moisture. They will spread steadily, especially if exposed to a burst of light every few years, so if they are in a border make sure that they are planted under shrubs that can be pruned hard every now and then, like cornus, roses, hazel or willows.

Next are the star-like wood anemones, *Anemone nemorosa*. They grow from a rhizome that creeps underground and, like primroses, do best in areas that are only lightly shaded, surviving the last few years of the coppice growth but bursting out in the years following a cut, providing pollen and nectar for bumble and honey bees. The flowers are white but fade to pink and will occasionally throw up a blue version and these have been used to breed blue hybrids such as 'Lismore Blue' and 'Robinsoniana' – although again, I prefer the native species. These anemones will colonise moist turf, and although I always think of them as woodland flowers, moisture seems to be the key to their success rather than the shade of trees.

Like primroses, bluebells survive the deep shade of the end of the coppice cycle in an attenuated form, flowering modestly where grass gives up the ghost and then, when a coppice is cut, the bluebells go berserk and spread themselves wildly before the grass gets a look in. In the garden it makes sense to grow bluebells in a situation that mimics coppicing, but it is a hopeless plant for any kind of border as it will completely take over once established and the only thing that will suppress it will smother anything else growing there as well. So it is all or nothing.

*(Previous pages, left) Wood anemones (***Anemone nemorosa***) are one of my favourite woodland flowers and spangle the floor of the Coppice with their white, star-like petals for a glorious week or so in April.*

(Previous pages, right) Bluebells are another essential woodland flower growing in the Coppice at Longmeadow. A carpet in woods is beautiful but in a border they are an invasive and obstinate weed.

Bluebells produce a toxin that helps them to fight off potential pests such as nematodes that would otherwise eat the bulbs, and attempts are being made to commercially extract this as a natural pesticide. In fact, nothing will eat bluebells because of their toxicity, which explains why you find them growing in great lakes in woods where all other undergrowth has been grazed off by deer or even cattle and sheep.

You can buy the Spanish bluebell (*Hyacinthoides hispanica*), which is less invasive, less slender and coarser in every way. It also hybridises with our native bluebell, so never grow the two together or else you will end up with a sum of the worst parts, an invasive, squat, less delicate and altogether less attractive flower.

EARLY CLEMATIS

Clematis must rank as one of the most popular plants in the British garden but unlike roses or tulips it hardly inspires great depths of romantic passion. Perhaps that is because it is so ubiquitous in its many forms but thanks be to that. Without clematis almost every garden would be a much duller place.

In April, the most common clematis of all, *Clematis montana*, is getting into its extravagantly floral stride – in the south at least – but the star of April for me is *Clematis alpina* in its various forms.

Clematis alpina 'Pamela Jackman' is completely untroubled by any amount of cold weather, coming, as the name indicates, from the Alps, and has been flowering blithely for weeks. The powdery mauve goat's ears petals are elegantly draped over the branches from the point when they are first coming into leaf until the foliage all but hides them. That's it then for the rest of the year, but it is enough.

We had a pair of *C.* 'Francis Rivis' growing up wigwams in the Jewel Garden. For a few years they liked this and flowered in a spectacular cone of delicate blue. But the woody stems inside the framework got bigger and denser and the hazel rods got more and more rotten and last year one blew over in a storm and killed the clematis. The moral of the story is to prune more ruthlessly than I did. Actually I now remember that the reason that I left it was that a thrush had made its nest in its bushy interior.

The time to prune early-flowering clematis is immediately after they flower to allow as long as possible for the summer's growth, that bears next year's crop of flowers, to ripen. Hacking them back any time after summer will result in no long-

term damage but will mean fewer flowers – or even none at all – the following spring.

I have one *Clematis macropetala*, growing up a tripod in the Walled Garden, with mauve, shredded petals. I coppiced this back to the ground this spring because it had become a tangle but I have done this before and know that it will regrow with enthusiasm and a little less convolution. It is one of my favourites, with a delicacy that the alpinas lack.

I like *Clematis armandii* but I don't grow it here, because it is too cold for it to survive – let alone thrive. But where it has some shelter – and we had a very happy armandii outside our back door in London in the 1980s – it is the best scented of all clematis and the blossomy white flowers are carried amongst large evergreen leaves.

I did plant a pink sort of *Clematis montana* in the Coppice to scramble over a hawthorn – the one original tree that was there when we came to Longmeadow in 1991 – but it scrambled so well that the tree was being killed for lack of light so I cut it off and it never recovered.

However, we have a borrowed *Clematis montana* that scrambles over the wall with our next door neighbour and hangs in a curtain of flower down in the propagating yard. Long may it do so.

Planting clematis

Soak the plant (in its pot) for 10 minutes in a bucket of water whilst you are preparing the planting hole.

Clematis like to have their roots in the shade in rich, well-watered soil. Prepare a hole at least twice the size of the container that the plant came in and add a bucket of compost and a handful of bonemeal, if you have any.

If you are planting your clematis to cover a wall, fence or tree make sure that the hole is at least 30cm (1ft) away from the base (including any footings) and angle the supporting cane back towards the wall.

Like roses, clematis need to be planted deep. Take the plant out of the pot, being careful not to disturb the rootball but gently freeing a few of the roots growing in ringlets at the base of the pot, and place it so that the soil level of the container is at least 2.5cm (1in) below the soil level of your hole. Fill back the topsoil to the junction

C. macropetala *combines freshness and exuberance with rich colour. This one is growing up a hazel wigwam in the Walled Garden.*

of plant and the potted compost and give it a good bucket of water. When this has drained, top up with a thick mulch of more compost.

If the base of the clematis is in the sun it is a good idea to put large pebbles or a tile or two around the roots to keep them cool and moist. Give it a full bucket of water once a week for the first few weeks and repeat if the weather is dry. If you are planting a clematis to climb an established tree you will have to water more frequently.

Whatever the type of clematis, once you have planted it, be bold and prune it down to a strong bud about 30cm (1ft) off the ground. This might seem drastic but it will ensure a strong, healthy framework to establish the plant and many more flowers in subsequent years.

EUPHORBIAS

Euphorbias have an astonishing green intensity that adds an electric charge to the garden as they emerge, energising everything – including me – around them.

The first to come into flower at Longmeadow is *Euphorbia amygdaloides* var. *robbiae*, which is in the Coppice. It has dark glossy green leaves with brilliant yellow flowers that it will carry into summer. It grows best in light shade and moist soil. The variety 'Purpurea' has a combination that I love of acid-green flowers and crimson stems and leaves which gives it real interest for the whole year and not just when the flowers are doing their star turn. These euphorbias can get mildew in summer if they become too hot and dry but a generous mulch every March will do a lot to combat this.

Euphorbia polychroma only grows to just over 30cm (1ft) making a flat clump of stunningly radiant yellow. However this display is rather short-lived so needs to be accompanied by something that will take over its role from the end of May. I grow it among my 'West Point' tulips in the Spring Garden, the two showing off disgracefully for a few weeks before giving up in exhaustion.

E. palustris is another that grows best in moist soil and also likes some shade. We have *E. palustris* 'Walenburg's Glorie' in the Jewel Garden and in early spring the plants are just producing new growth which will romp away once they get going and need support if they are not to flop over all their neighbours. The foliage turns various shades of orange in autumn with 'Walenburg's Glorie' notable for its exceptionally brilliant autumnal hues.

In early and mid-spring the best euphorbia for a border in full sun is *E. characias*. There are a number of excellent varieties to choose from and all share the characteristic black eye to each electric lime-green flower and all have the most shrub-like structure of any hardy euphorbia and so are useful as structural plants within a border. We have *E. c.* subsp. *wulfenii* 'John Tomlinson' in both the Jewel and Walled Gardens and the flower spikes last from late April right up to midsummer. In the first year it will produce spires of grey leaves without any flowers. This is because although the plant is perennial, the flower spikes are biennial, so they will follow the next spring. Obviously once established it will have both kinds simultaneously so you will always have a display. Do not worry if it seems to be toppling over under its own weight – it has a tendency to start its growth bowed over but it gradually straightens up to become a long mane about 1.2m (4ft) tall, of inverted glaucous leaves topped by flowerheads. When the flowering has stopped, the stems should be cut back as near to the ground as possible.

Just one word of warning: all euphorbias produce a milky sap that can cause an allergic reaction to the skin, especially in sunlight. So wear gloves when you cut them back and if you do get any sap on your skin, wash it off immediately.

EARLY HERBACEOUS PERENNIALS

The extra light that is pouring into every day is the important trigger for early-season plants. There are two main groups of garden plants strikingly affected by this: the early annuals and the early perennials that mainly choose to grow in woodland conditions where the midsummer canopy of tree leaves acts as an umbrella blocking light and moisture. So they have evolved to grow fast as soon as the spring days lengthen, to flower before the canopy closes above them.

Herbaceous plants have evolved to die back completely in the dormant season to protect themselves against cold, although the roots remain alive and healthy, throwing up new, non-woody growth in the spring. Because its season is so short, the plant has to grow with tremendous vigour in order to flower and set seed before autumn. Hence the dramatic transformation in the herbaceous border in the months from April to the end of June.

Most of our early herbaceous plants are in the Spring Garden. The first to appear is pulmonaria or lungwort. We have *Pulmonaria angustifolia*, which has violet flowers growing out from chocolate calyxes that look remarkably like dried up seedheads until a very undried-up flower emerges. Pulmonaria flowers often start pinkish, turning bluer as they develop but there are truly pink, almost apricot pulmonarias, and *P. rubra* is a soft red, as is 'Margery Fish'. We also have *Pulmonaria officinalis* 'Sissinghurst White', which has icy white flowers and leaves mottled and spotted with white, as though someone has spattered milk over them. These leaves form in summer after the flowers have finished, so the time to cut them back is either immediately after or before flowering according to the extent of mildew, to which they can be prone.

Another famously blue flower in the season of yellows is *Corydalis flexuosa*. This is one of my favourites, both for the incredible china blue of its flowers and its glaucous leaves that have a trembling fragility, yet it is a common enough plant and not difficult to grow. It is a close relation of the poppy and dicentra, and also the common fumitory – the latter is a regular but welcome weed in the Walled Garden. Its blue flowers will retain a freshness and intensity of colour if they are grown in cool shade – but that suits the plant anyway. There are a number of strains you can buy, like 'China Blue', and 'Père David', which has bronzish leaves. Like pulmonarias, corydalis thrive in cool, light shade so to keep them flowering shade them from midday sun.

Dicentra spectabilis, or bleeding heart (which has recently been subject to a name change to the ugly *Lamprocapnos spectabilis*), has similar leaves to corydalis and feels as like an English cottage garden plant as any, but the best known originally come from China. *Dicentra spectabilis* has little pink heart-shaped flowers with a tassel hanging in a neat row from its arching stem. The white version, *D. s.* 'Alba', is also lovely and contrasts better with the glaucous leaves. Like the corydalis, dicentra is such an essentially spring plant and I cannot imagine the season without it.

Heuchera, tiarella and tellima all share the same growth pattern of a low mound of leaf beneath tall, fragile shimmers of flowering spikes that are individually hardly more than dots of colour but collectively work really well, especially amongst the dappled shade of shrubs. None of these shouts from a border but for all their modesty they have a seductiveness that captivates.

Less ethereal are the early geraniums, which in my garden are represented by dark purple *Geranium phaeum*, the magenta *G. maccrorhizum*, and the pretty pink of *G. endressii*. These early geraniums will grow in sun or shade, cover the ground well and, perhaps most usefully, take being hacked back around midsummer to let light and air in and grow back for a late-season flowering.

Increasing your stock

- This is an excellent time to move herbaceous plants and to create new plants from old. If the plant is relatively small this will mean cutting it in two whilst it is in the soil and digging up one of the halves. If it is a larger plant it is a good idea to dig up the whole thing, place it on the surface of the soil and cut pieces – either in segments like a slice of cake, or as though you were reducing a circle to make a square by slicing off the outsides. Use these outside sections for the new plants and discard the centre. Water the new pieces in well and they will grow away as though nothing had happened.

- Herbaceous plants are voracious feeders and the better and richer your soil is, the better the display. However do not feed your plants unless they are sickly. This only leads to unnaturally lush growth that is especially prone to attack by pests and disease. Always feed the soil not the plant.

SUMMER BULBS

Summer-flowering bulbs have mostly evolved to survive winter cold and drought and they tend to come from parts of the globe where summers are warm and moist. So it is best to plant summer bulbs in spring as the soil begins to warm up in tune with the rhythm of their growth. The exceptions to this rule are alliums, which should be planted between October and January and in any event span the two seasons of spring and summer.

Lilies are perhaps the most openly sensuous of all summer bulbs and although they are strongly associated with florist flowers, many are easy to grow in the garden. All lilies have bulbs with open scales like a half-opened artichoke. Most are essentially woodland plants, and like cool, shady roots and sunshine on their faces. Many grow best in slightly acidic soil although the Madonna lily (*L. candidum*) is the one exception that does best in alkaline conditions. Some lilies such as *L. canadense* or *L. nepalense* are stoloniferous so they quickly colonise an area if the soil is loose and fluffy with leaf mould. In fact, leaf mould is the key to lilies – they love the loose root run it encourages. But be warned that *L. nepalense* likes to be dry in winter and moist in summer – a combination we find hard to produce at Longmeadow.

Lilium regale is one of the easiest to grow and one of the best. It is very hardy, copes equally well with acid or alkaline soil but does need good drainage. I always plant the bulbs with a good scoop of horticultural grit beneath them so they never sit directly on damp soil. However *Lilium henryi* loves the soil of the Damp Garden, which regularly floods, and its orange Turk's-cap flowers in late summer are a joy.

Lilies always do very well in pots as long as you add plenty of leaf mould to the potting compost. The key to lilies in pots is to give them enough room. Don't cram half a dozen bulbs into a small container. Give them plenty of space – about three bulbs for a 5-litre or bigger pot – and they will grow strongly and produce masses of flowers over a long period.

The Cape hyacinth, galtonia, is a good border bulb as long as it has plenty of sun (which means not allowing neighbours to shade it) and the soil does not dry out. *G. candicans* is 90–150cm (3–5ft) tall with white bells of up to 30 flowers and is one of the stars of the Damp Garden. *G. viridiflora* has pale green flowers and is hardier although, given our recent winters, it is best to lift the bulbs and store them in pots over winter.

Cardiocrinum is a relative of the lily and superbly dramatic with *Cardiocrinum giganteum* able to grow to 4m (13ft) tall with huge glossy leaves and a flower spike of white trumpets. It needs a moisture-retentive (but not wet) soil and – more difficult to control – a damp atmosphere. Dry summers do not suit it at all and I lost half a dozen precious (and very expensive) plants in the summer of 2011. The bulbs die after flowering and the offsets may take another three or four years to produce a flower, so new bulbs should be planted each year, keeping the tips just below the surface.

Although much less exotic, I adore crocosmia and it is one of the easiest summer bulbs to grow. They like a good rich soil with plenty of organic matter to retain moisture and then lots of sunshine. *Crocosmia* 'Lucifer' is perhaps the best-known and is fabulous. There are other really good cultivars, such as the later, more orange, more delicate and longer lasting *C.* × *crocosmiiflora* 'Emily McKenzie', 'Bressingham Blaze' which is bright red and *C.* × *c.* 'Citronella' which is a delicate yellow. The secret of growing crocosmia is to plant the tiny corms deep – at least 15cm (6in) – so that they do not dry out.

Finally, eucomis, another green-flowered bulb whose pineapple flowerhead looks almost tropical but which is also from South Africa. These are surprisingly hardy and thrive with plenty of moisture (and some shade) in summer but need to be dry and frost-free over winter – so the dormant bulbs should definitely be lifted in winter and replanted in late spring. *Eucomis comosa* is the most commonly grown but *E. bicolor* is better with lovely pale green flowers. It works particularly well planted with hostas in the Damp Garden.

TULIPS

Whilst I love the delicacy of many spring flowers such as primroses or the very early yellow roses, there is something uniquely and exuberantly joyful about the intensity of tulips. They are the first real blaze of colour of the year and are the most silkily sensuous of all flowers. Although the best time to plant tulips is November, now is the moment to relish their wonderfully brilliant colour and choose your favourites to grow in your garden next year.

Tulips grow wonderfully well as part of a mixed border, in long grass or in pots. Rather than seen as a display in their own right – which of course they can perfectly well be – I think that they work best in the garden when mingled and mixed with other spring plants.

The very first tulips to appear in my garden are the yellow species tulip, *Tulipa sylvestris*. We planted these 10 years ago at the base of a hornbeam hedge, which 10 years on, has become much, much bigger and almost swamps the tulips. Nevertheless they grow back happily each spring, flowering in mid- to late March with buttercup-yellow loose petals on loose, floppy stems – despite the bulbs growing right amongst the hornbeam roots on very wet soil. I suspect that the sunny aspect helps and that the hedge is taking up most of the moisture so stopping the bulbs from rotting. Last year I planted another batch in grass between the Coppice and Soft Fruit Garden and they all flowered well. But tulips always look good in grass – I have 'Lady Jane' growing under apple trees in the Writing Garden, but it does mean that the grass cannot be trimmed until July when the tulip foliage has completely died back.

After the species tulips, which tend to flower first, the earliest to bloom in my garden are 'Abu Hassan', which has a lovely burnished copper tone to it; 'Negrita', a plum-coloured flower; and 'Queen of Sheba'. The first two are from the Triumph group, which are bred to have long straight stems that make them good for cutting.

'Lady Jane' tulips growing in grass under the apple trees of the Writing Garden

The sumptuous 'Queen of Sheba' is always the first of the lily-flowered group, along with 'Prinses Irene', which has a chocolate stem and orange petals flushed with pink and streaked with plum and a touch of green.

Parrot tulips have frilly, fuzzy edges and always look to me as though they are caught in the middle of exploding out a rich splash of satiny colour. I have 'Black Parrot' which is actually a deep purple rather than black and 'Orange Parrot', which is very much as its name suggests and flowers to the end of May. 'Rococo' is a violent splash of crimson and 'Flaming Parrot' has pale yellow flowers flamed garishly with a raspberry streak.

The lily-flowered group is less flamboyant but I think my favourite. As well as 'Queen of Sheba', it includes the primrose-yellow 'West Point' which is as good as any of a very good bunch, which I have mingling amongst the forget-me-nots in the Spring Garden alongside Solomon's seal and euphorbias. From the same group is the pure white 'White Triumphator' which I have flanking the path under the pleached limes along with pale yellow 'Nicholas Heyek'. The last to flower and the last to stop flowering in my garden is 'Queen of Night' which is in the Herb Borders along with 'Recreado' – another very good, dark tulip.

As with all bulbs, do not cut tulip leaves back after flowering as they need to feed and create the new bulb for next year's flowers. However, the whole plant, bulb, leaves and all, can be dug up and stored in racks or replanted in a nursery bed. When the leaves have died back naturally the bulbs can be stored until autumn.

Tulips will only produce one good flowering bulb each year together with a number of smaller bulbils. These may take a few years to develop a bulb big enough to flower, which is why tulips left in the ground tend to get smaller and more numerous each year. It is best to top up the display with some fresh bulbs each year.

The fungus *Botrytis tulipae* causes tulip fire. I speak from bitter experience because all our tulips in the Jewel Garden were afflicted with this in 2006 and 2007. Its first symptoms are pin-sized bleached marks on the flowers, then withered, distorted leaves that may finally rot and be covered in a fungal mould. Any affected plants should be lifted and burnt, and no tulips should be planted in the same spot for at least three years.

We grow lots of tulips in borders – as with these 'Recreado' in the Herb Garden.

LETTUCE

The more that I grow vegetables – and I have been doing so for over 40 years now – the more I value fresh salad crops. Of course I love the whole range of vegetables that my garden and greenhouse can provide, but absolutely nothing tops a simple salad eaten minutes after gathering it from the garden. When I am home I try to eat a home-grown salad every single day of the year. And when I am away, no restaurant salad ever tastes as good.

There are dozens of leaves as well as tomatoes, cucumbers, beetroot, radish, onions and fennel that you can grow to eat as a salad but the mainstay, the core of salad production, are lettuces. I like to be able to go out and gather lots of lettuce at any time (although between November and March this can be tricky) and at any one time we have four or five different types of lettuce growing at Longmeadow, with more at various stages in the greenhouse and coldframes.

Lettuce varieties and types

One of the big advantages of growing lettuce at home is that the range available to the gardener is very much greater than to the shopper.

The oldest and most familiar variety is butterhead, which has rosettes of soft, cabbage-like leaves. As a rule, butterhead lettuce are a little too waxy and soft for me – but they do grow well in cold weather so can provide fresh leaves in winter, and good butterheads like 'Tom Thumb' have a lovely delicacy and freshness. 'All The Year Round' is, as the name implies, hardy and adaptable enough to crop most of the year and whilst not the best you can grow, a lot better than almost anything you can buy – especially in that spring gap when there is little else in the garden. 'Valdor' is an overwintering type I grow, sowing the seeds in August for harvesting from Christmas onwards. Butterheads store poorly so should be cut and eaten on the same day.

Various lettuce seedlings growing in seed trays in the greenhouse. These are ready to be pricked out into plugs and grown on before planting out as vigorous, slug-resistant transplants.

LETTUCE
RED SALAD BOWL
28.5.11
LETTUCE
MARVEL OF FOUR SEASONS
28.5.11

I love a good crisp cos lettuce. 'Little Gem' will do well from an early sowing and certainly is always worth finding room for in the garden. A home-grown one will astonish you with its freshness and taste compared to the supermarket version. 'Lobjoit's Green Cos' is even better – much bigger, rather slower to develop and with a shorter season – but one of my favourites. It needs plenty of water to heart up well and if allowed to become dry the outer leaves will rot. 'Rouge d'Hiver' is a cos that will grow in cooler conditions (although not over winter) and 'Paris Island Cos' is another one that I grow and enjoy. 'Winter Density' is a cos hybrid that will stay small over the winter months and grow in spring for an early harvest. You can get red cos too, which adds variety if not any extra taste.

Cos lettuces are higher in vitamin C and beta carotene than the ubiquitous iceberg, which, when bought, usually has no taste and is grown solely for its admirable crispness. But it's worth growing an iceberg to see what it can taste like when it's not factory farmed. 'Webb's Wonderful' is the best-known variety, and 'Chou de Naples' is the parent of most modern iceberg lettuces with the virtue of being slow to bolt.

If a tasty iceberg is a rare experience, loose-leaf or salad bowl is nearly always good and can be picked leaf by leaf or cut flush to the ground and left to regrow for one (and usually two) further cuttings. I grow red and green oak-leaf or salad bowl varieties (they come with various proprietary names). 'Lollo Rosso' is one of the best known but is a dreadful lettuce. Try 'Red Salad Bowl' instead. It is far nicer and looks fantastic. Red lettuces grow slower than green ones and tend to be a little bitter, which I like. They are also less likely to be eaten by slugs than green leaves.

A 'Little Gem' lettuce ready for harvest. I grow lots of lettuce and always have enough to harvest them very young whilst they are exceptionally tender.

Growing lettuce

There are two ways to grow every type of lettuce. The first and traditional method is to sow them directly into the ground where they are to grow (including in any kind of container), in drills or broadcast. You then thin them carefully as soon as they are large enough to handle, thin again a few weeks later (eating meltingly tender thinnings, roots and all) so that you are left with a row of maturing plants about 8–23cm (3–9in) apart. The advantages of this are that the roots of the plants are not disturbed more than necessary and you need no potting compost, seed trays, plugs, greenhouses or other paraphernalia. As long as the growing medium is well drained and quite rich – I always add and lightly rake in 3cm (1in) of fresh garden compost before sowing or planting out salad crops – they should grow well.

The disadvantage is that they are susceptible to snails and slugs, especially at the young seedling stage and particularly at this time of year when growth can slow to a standstill during cold weather. Handling is also no less if you sow them in drills because the thinning process is fiddly and essential. They must also be watered and kept weed-free. However, handling is arguably just another word for gardening and enjoyable in itself. Personally I only sow direct when using a mixture of seed such as 'Saladesi' or 'Saladini', which I broadcast and do not thin except to eat.

I find it just as easy and much more controllable to sow the seed in plugs or seed trays, grow them into reasonable-sized seedlings and then plant out at 23cm (9in) spacing when big enough to withstand slug or snail attack. This system depends upon some kind of cover. A greenhouse is best but coldframes are very good and a porch or spare windowsill is perfectly workable.

Most lettuces take a month to six weeks to grow sufficiently large to eat, and will last for 8–10 weeks after sowing. To provide a constant supply of salad leaves, sow a few seeds every 10–14 days from mid-February through to late September so as one small harvest is used up another takes its place.

Lettuces germinate at surprisingly low temperatures and many will fail to germinate once the soil temperature rises above 25°C (77°F). At the same high temperatures, mature plants will almost overnight develop flower stems and run to seed. This makes the leaves bitter. There are ways around this. If you are sowing directly into the soil, water the drill before sowing to cool the soil down. Sow in the afternoon so that the vital germination phase coincides with the cool of night. Also sow into a shaded part of the midsummer vegetable garden.

Lettuce pests and diseases

Slugs and snails love a juicy young lettuce more than anything else. The best organic defence against slugs is to grow the seeds in sheltered, slug-free conditions, harden them off for at least 10 days and then plant out into their final growing position 15–23cm (6–9in) apart. They then tend to grow strongly enough to resist slugs at the critical seedling stage. However, I always wash summer lettuces carefully as they are full of slugs.

Lettuce root aphids attack lettuces later in the season if they have become too dry. The first sign is when the plants visibly wilt and then die. When pulled up the roots have a white, powdery waxy coating discharged by little yellow aphids (*Pemphigus bursaurius*) that eat the roots. Do not grow lettuces in the same soil for a full year.

Downy mildew will turn outer leaves yellow and then pale brown and the undersides of the leaves develop a nasty downy growth. This fungal disease is encouraged by humid conditions, and the only solution is to remove all affected leaves and thin the batch to increase ventilation.

Lettuce grey mould is another fungus (*Botrytis cinerea*) that produces fluffy grey mould on the leaves. Sometimes this is first identified by a slimy brown rot on the stem and the whole plant collapses. Good ventilation and healthy plants are the best corrective.

MIZUNA, MIBUNA, LAND CRESS & LAMB'S LETTUCE

I have grouped these crops together because they share similar growing conditions, doing best in cool weather, and I use them as spring and autumn salad leaves.

Mizuna and mibuna are both members of the mustard family and from Japan. Mizuna has finely serrated leaves and will grow much like salad rocket. However, each plant can become more substantial than rocket, and the longer stems are crispier and better to eat. It is best harvested by cutting all the leaves with a knife, allowing at least one (and often two or three) new flushes of foliage.

Mibuna has long, strap-like leaves that are milder than mizuna and it is a little less hardy too – but ideal for growing in a greenhouse or tunnel over winter. It makes a rather handsome plant with a fountain of leaves and is a very useful winter salad standby. Both mizuna and mibuna will tolerate drier conditions than rocket.

Land cress (or American land cress) looks very similar to salad rocket but is much hotter and more peppery and has a distinct similarity to water cress. It loves rich soil with lots and lots of water and is very hardy, but it's very susceptible to both flea beetle and bolting when the weather turns warmer.

Lamb's lettuce

Lamb's lettuce (or corn salad, and the Americans only know it as maché) can be much trickier to grow than any of these leaves I've mentioned. It is a hardy annual but is relatively slow to germinate. Sow it in situ, either broadcast (apparently it was the tradition in eastern Europe to broadcast onion beds with corn salad as a companion crop) or in shallow drills, although I often scatter seed in a seed tray and prick out the seedlings before transplanting outside. It grows slowly, taking at least a full three months from sowing to the first harvest. It likes cool, moist conditions. In fact, wherever land cress is happy, corn salad will also thrive.

Thin the plants to 15cm (6in) spacing and cut the rosette of small leaves with a knife. Each plant will provide at least one further harvest before going to seed. If you have a cool, damp corner leave a couple of plants to go to seed there and they will regenerate themselves more or less indefinitely.

LEGUMES

With intelligent use of nitrogen-fixing legumes, you never need to add nitrogen to the soil. No nitrates means an instant reduction of river pollution, ill-health, and the array of pests and diseases caused by the over-production of soft, sappy growth that too much nitrogen invariably promotes.

This approach means that you put your efforts into creating a healthy soil that feeds its plants what they require rather than pumping sick earth with chemical resuscitation to get you through to the next crisis. This is something that organic farmers have always known – and all farmers have always been organic until, historically, the day before yesterday – and that any vegetable gardener has always practised as part of their crop rotation. Legumes first, brassicas second and roots third, before manuring the ground again and going back to legumes.

If you do not grow peas or beans then there are green manures (such as alfalfa, field beans, fenugreek, winter tares and trefoil as well as the various clovers) which will all fix nitrogen, suppress weeds and provide organic material to be dug into the ground directly or cut and added to the compost heap.

PEAS

Peas should not be sown into wet ground nor when the soil is cold to touch, otherwise they have a tendency to rot. I have found that a double row 23cm (9in) wide with each 10–12cm (4–5in) apart works best. Either prepare the soil to a fine tilth and simply push the peas into the ground, or draw a shallow drill with a wide hoe, place the peas along it, and then rake the soil back over them. There should be room between each of these rows to walk and pick the pods, which in practice means at least 1m (3ft).

Peas need support to make picking easier. I like to use pea sticks – which can be any kind of twiggy brushwood or prunings, although hazel is best. Netting or chicken wire works very well if supported by bamboo canes woven into the wire and pushed into the ground. A short variety like 'Kelvedon Wonder' will only grow to 90–120cm (3–4ft) but a taller variety like 'Alderman' can reach over 2m (6ft) so needs substantial support if the pea bines are not to topple.

Whilst they should not dry out as they are growing the important time to water them is once they flower. This will encourage pea-filled pods. Pick the pods as soon as they are large enough to open, which will encourage more pods to develop. They soon shrivel up in hot weather.

I like 'Alderman' for its old-fashioned extravagance of height and unsurpassable flavour. I also grow 'Hurst Green Shaft' and 'Carouby de Maussane' and have often grown 'Feltham First' when I made a very early sowing. 'Kelvedon Wonder' is ideal for small gardens as it is a dwarf variety with good flavour.

Mange tout and sugar-snap peas are eaten pods and all. The former never really develop proper peas and retain flat pods, and the latter do so slowly, which means that if you do not pick them regularly you can harvest the maturing peas and eat them as a normal variety. Both lack the hard wall to the pod which conventional wrinkled peas have. There is a lot to be said for growing sugar-snap peas if you have limited space and perhaps limited patience for shelling peas.

(Previous page) Tying a hazel wigwam for climbing beans. I cut these bean sticks from the Coppice and regard them as being almost as decorative as the beans themselves.

BROAD BEANS

Broad beans are delightfully straightforward to grow. They are very hardy and can be sown as early as October for a late spring harvest, although I also sow on the first day in spring that the soil can be worked, with another couple of sowings thereafter until June. Like peas, a double row 30–45cm (12–18in) apart is best, with the beans planted 23cm (9in) apart down the row. I use my ubiquitous scaffolding plank and crawl along it whilst pushing the beans into the soil down its length. Like with peas, 90cm (3ft) is necessary between each double row to allow room to pick.

Although broad beans are free-standing, they will need support to stop them being blown over. The best way to do this is to stick a cane firmly every few feet along both sides of each row and to stretch string between the canes so that the beans are trussed upright.

Broad beans are best when small and will soon grow too fast to keep up with. Once larger than a fingernail the beans are better blanched and skinned before cooking.

Although unsightly, blackfly do little harm. They are attracted to the soft growth at the top of plants in midsummer, so pinch these off once you see blackflies collecting.

Varieties: 'Aquadulce', 'Red Epicure' and 'Express'

(Following pages, left) The brushwood tops left after cutting bean sticks make excellent support for peas so nothing is wasted from the harvest of coppice hazel..

(Following pages, right) Broad bean seeds, ripple-skinned yet pebble smooth, are the most tactile of all seeds and certainly none is as easy to sow.

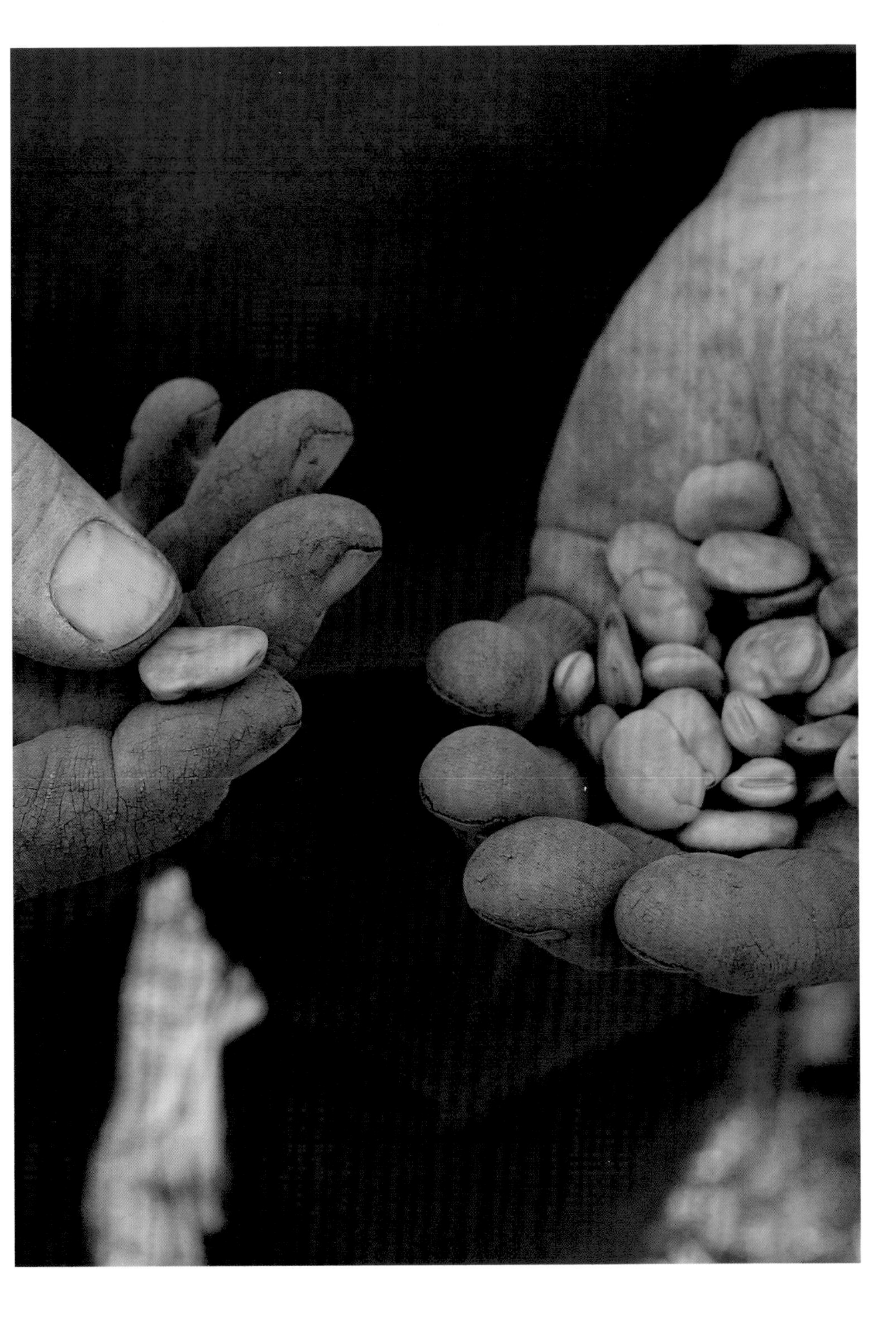

FRENCH BEANS

French beans (which are not French at all, coming, as do runner beans, from Central and South America) are *Phaseolus vulgaris*, and include haricot beans, kidney beans, dwarf, pole, yellow, green, purple and blotchy ones. Unlike peas or broad beans they are frost tender and it is a mistake to sow them too early. The great secret of sowing French or runner beans is to wait until the nights are warm – even if that means not sowing until mid-July.

Dwarf beans

I sow dwarf beans in blocks rather than rows, a hand span apart, pushing each bean into the finger-soft soil. Slugs and snails are very fond of the young plants and can reduce them to stumps overnight. For this reason I always grow some in plugs and plant them out when they are established, at which point they are left alone.

Varieties: 'Roquefort', 'Purple Queen', 'Annabel' and 'Purple Teepee'

Climbing French beans

These are always superbly decorative with a choice of green, yellow or purple beans, and add essential height to the vegetable garden. By the same token they are ideal for limited spaces and grow well in a large container. Like runner beans, they need rich, water-retentive soil so I dig a pit and add plenty of organic material before covering it over with soil. I then build a teepee above the hole from four good poles lashed firmly together at the top, and plant a couple of beans at each corner.

Varieties: 'Burro d'Ingenoli', 'Blue Lake', 'Hunter' and 'Blauhilde'

RUNNER BEANS

Runner beans are one of the favourite vegetables for British gardeners, decorating every allotment up and down the country. I grow them in exactly the same way as climbing French beans. They thrive in warm, wet weather and will tolerate a surprising amount of shade as long as they are not cold and dry. Harvest the beans when they are no more than 30cm (12in) long and be careful not to overcook.

Varieties: 'Desiree', 'Streamline' and 'Painted Lady'

POTATOES

One of my first jobs in April is to plant potatoes. But, as with almost everything in the garden, it is wrong to follow the calendar slavishly. I have planted potatoes as late as June and harvested a good crop from them in October. Whenever you do it, the soil must be dry and if not exactly warm then not cold to touch. Farmers of old would drop their trousers and sit bare-cheeked on the soil to test the temperature. Perhaps that might not go down too well on the allotment or in your veg patch, but the principle is good.

There are more than 400 varieties of potato divided into three groups: first earlies, second earlies and maincrop. First earlies mature after about 100 days, second earlies between 110–120 days and maincrop after 130 days.

Back in 2006 we did a trial on *Gardeners' World* to see if chitting – letting seed potatoes sprout before planting – helped, and we found that it clearly improved the yield of earlies but made no difference to maincrop harvests.

Potato planting is one of those things that must be done with rhythm and concentration, but in the end it is simple enough: stick a tuber in the ground, cover it up and wait for it to grow.

Ground for potatoes should be well dug and richly manured, preferably a few months before planting. If that has not been possible, add some garden compost along the bottom of the planting drill or individual hole. Make a drill or shallow trench about 15cm (6in) deep and put the potatoes in the bottom at about 30cm (12in) spacing. The wider the spacing, the bigger the potatoes, therefore, new potatoes tend to be sown closer together and maincrop – especially baking varieties – wider apart. Draw the sides of the drill back over the line of seeds to make a low mound. A mattock is the ideal tool. Each row wants to be about 90cm (3ft) apart to allow room for earthing up, which means drawing soil up over the emerging foliage to protect it from late frost and to make sure that the tubers have a decent layer of covering soil.

Potatoes are very good for planting in new ground that you want to use later in the year, as they open up the soil and their thick top growth stops any weeds growing. Rather than dig it over, all you need to do is make the trench and fill the bottom with

compost, putting the seeds directly onto that before covering them over. I have grown a perfectly good crop on fresh grassland by lifting turves along the line of each trench, placing the seed directly onto the soil and replacing the turves upside down on top of them.

With anything other than a large garden it makes sense to stick to earlies for a few delicious meals or to grow a really special maincrop variety like 'Pink Fir Apple' or 'Ratte'. Best of all, whatever the variety, is the taste of your own potatoes grown in home soil, freshly dug and taken straight to the kitchen then boiled in their scrubbed skins and eaten still steaming, sprinkled generously with good olive oil or butter and sea salt. That is what gardening is all about.

Varieties:

FIRST EARLIES (100 days)
- 'Rocket'
- 'Duke of York'
- 'Belle de Fontenay'

SECOND EARLIES (120 days)
- 'Roseval'
- 'Charlotte'
- 'Maris Peer'

MAINCROP (130 days)
- 'Pink Fir Apple'
- 'Sante'
- 'Remarka'

Maincrop potatoes drying in the afternoon sun before being collected up for storage. If kept cool and completely in the dark they should keep well into spring.

I always sow a drill of rocket or radish between rows of newly planted potatoes. This 'catch' crop will grow quickly and be harvested before the potatoes need earthing up or shade them out.

RADISH

Radishes have been a mainstay of salads for thousands of years. These roots were first domesticated in the Mediterranean, cultivated by the Egyptians nearly 5,000 years ago, and then introduced to Britain by the Romans nearly 2,000 years ago.

Clearly their hot, peppery flavour is popular across all cultures but another reason for their ubiquity is that they are probably the easiest vegetable to grow. Just sprinkle the seed thinly in a shallow drill 1cm (½in) deep and they will appear within a week and be ready to start harvesting a few weeks later. It is important to thin them so that they are at least 2.5cm (1in) apart which allows each root to become juicy and swollen before it gets woody. The thinnings can be eaten – leaves, roots and all – when they are the size of peas, although the perfect size for a radish is the circumference of a two-penny piece. Fresh radishes in early May eaten with salted butter are one of the seasonal treats of spring.

They will run to seed in warm or very dry weather so sow as soon as the soil is warm enough in spring and repeat a sowing every two weeks thereafter. Water regularly, but too much moisture will produce extra foliage and split roots.

I sow radish in the same drill as my parsnips (as the latter are slow to germinate): the radish mark the drills and are harvested before they compete with the emerging parsnips.

Varieties: 'Cherry Belle', 'Saxa', 'Scarlet Globe', 'Flamboyant' and 'French Breakfast'

ROCKET

This is the time of year when rocket is at its best, combining the familiar peppery flavour with a lovely, much less-known buttery texture that you very rarely come across in shops or restaurants. There are two kinds of edible rocket, the Mediterranean salad rocket (*Eruca sativa*, and also known as arugula) and wild rocket (*Diplotaxis*). Salad rocket has rounded leaves with varying degrees of indentation and at best almost melts in the mouth, whereas wild rocket is deeply indented and tougher.

Both are delicious – but if I could grow only one it would be salad rocket as it is easier to germinate and much faster to grow. After radish it is the easiest of all seeds to germinate. Chuck them out onto any old soil or compost and they will grow. But the crucial thing is to thin boldly and transplant them early. Rocket seeds are biggish and reasonably easy to handle so I now sprinkle two or three per plug, wait until at least two germinate successfully and then ruthlessly thin down to one healthy seedling. These are grown on and hardened off until they are about 5–8cm (2–3in) tall, before planting out at 23cm (9in) spacing. You can sow them direct but they will still need fierce thinning. This will seem ridiculously generous with space but closer spacing does not result in any greater harvest and the widely spaced plants receive lots of nutrition and water so become healthier and longer lasting.

Rocket hates hot or dry weather and will run to seed quickly – but rich soil, some shade and plenty of water will delay that process. However, for another couple of months, it is the easiest and fastest salad crop to grow and will do just as well in a pot or window box as in the vegetable garden.

Compost

Compost seems to me to be the perfect embodiment of what any good gardener is trying to do. There you have heaps of waste, which by definition is the least valuable thing you might have, metamorphosed into the single most useful thing in any garden. It is a deeply satisfying alchemy.

I never grow tired of this magic. It is, I guess, the same kind of satisfaction that you get from making bread or wine, where the possibility of disaster is real and yet experience and trust in the process allow that fear to become a frisson rather than an anxiety. It tends to work.

But for all this I also know that many gardeners still find it a bewildering and frustrating process that never quite goes right enough for them. Instead of a delicious crumbly mixture fragrant of a woodland floor in spring, a slightly dubious sludge emerges spotted with undecomposed objects like the odd carrot, envelope or lemon. I assure you that this happens to the best and most compost-sure of us. But any compost position can be rectified.

It is worth understanding what is going on in that mysterious heap. Compost is not the product of decay but digestion. It is a measure of the extremely efficient digestive systems of billions of hungry creatures, albeit mostly extremely small ones. We see beetles, slugs and above all worms – the red brandling, *Eisenia fetida*, that will appear in any well-made heap by the thousand – but most of the work is done far, far beyond the reach of the microscope, let alone the unaided eye. The compost maker simply has to create the best possible circumstances for that digestive process to happen.

This means that instead of seeing a compost heap as a place where garden and kitchen waste rots into a kind of socially acceptable form of manure, it is more appropriate to regard it like a sour dough starter or yeast, packed with the most intricate forms of life imaginable, condensed into a marvellously manageable form. This then goes back into your soil not so much to fertilise it as to recharge and re-energise the life that is already in it.

Nothing nurtures that web of subterranean life better than compost. By adding small amounts of compost you are topping up the ecosystem that converts plant and

(Following page) Every scrap of garden waste is recycled back into the soil via the compost heap. We put it all into a holding bay, then shred it before adding it to the first pile.

animal debris in the soil to humus. Humus is the ideal soil because it holds nutrients and water within easy reach of plant roots and gives it that lovely spongy, crumbly structure that both plants and the creatures of the soil like best. It is therefore a virtuous circle. Good soil is constantly renewed and made better. But the key to this is the living population in the soil and the main purpose of compost is to nurture that ecosystem.

Making compost is easy. It wants to happen. Anything that has ever had life in any form is grist for the compost mill, be it massive oak or the discarded leaves of a lettuce. Accumulate a pile of organic material in the corner of your garden – and this could include paper, hair, wood and wool as well as the more conventional grass mowings, kitchen waste and weeds – leave it for two or three years and you will have perfect compost. Guaranteed. If you have the space this is perhaps the best and easiest method of making it and I noticed at Ninfa, that glorious garden south of Rome, that there they have up to a dozen compost heaps like this all over the garden rather than one main composting area. But the average back garden cannot accommodate this. I know that the two acres of this garden cannot.

So most of the human energy in making compost is taken in speeding the process up. The faster you can turn waste around the less space you need to tie up. On balance, I find that it is reasonable to expect compost to take anything from 4–10 months to make.

At Longmeadow I have evolved a system that works for us. It is not a blueprint, just an example of what works in our kind of set-up. We have four bays made of corrugated iron. There is also a small holding bay and everything gets dumped in here except exceptionally prickly prunings, which are burnt.

Every Friday morning the contents of this first bay are shredded and added to the second bay. What cannot be shredded – grass clippings, old lettuce, tea bags and what-have-you – is forked directly onto this second heap, but the more we can shred, chop, mow or crush the material, the quicker it will break down.

The second bay is the most active one and is topped up over a period of about a month. The ideal is to make it a mixture of green and brown material. Another description of this is a mixture of green material that is high in nitrogen – like grass and lettuce leaves – and brown stuff with much more carbon – like cardboard or dry herbaceous growth at the end of summer.

The bacterial activity that makes compost seems to work best when the ratio is about 25:1 carbon to nitrogen. This is not an exact science but is the main reason why a compost heap fails. Damp grass clippings, for example, are about 50:50. Garden soil is about 12:1 but sawdust is about 400:1. It follows that there needs to be a mixture of browns and greens to create a balanced heap with the emphasis on plenty of browns.

That mixture does not need to be added in layers. Just chuck it all in. But most waste material from a garden and household tends to be high in nitrogen and will benefit from the addition of more brown material. If you have a large cardboard box I suggest using it as a container to store very carbon-rich material. Leave it outside exposed to the rain as this accelerates its decomposition. Newspaper is better recycled than composted as the inks seem to inhibit the composting process.

When this second bay is full I turn it into the third and when that is filled up it goes into a fourth and then a fifth and final compartment. Turning is vital because it lets in oxygen to the heap. Without air a compost heap either turns into an anaerobic evil-smelling slime or a dry brick of undecomposed matter. Oxygen is the key. The reason for this is that most of the bacteria that make your compost need oxygen to survive. When you turn it you are literally breathing life into it. If you have a small container all you have to do is empty it and put it all back in again. If you have smallish cardboard boxes, don't flatten them and put them in as layers but scrunch them up so they create air pockets. Air is the single most important element of good, quick compost.

Water is also important. If you have the right mix of brown to green material it is easy to find that your heap becomes rather dry. Keep it moist, putting a sprinkler onto it if need be. If it becomes sodden it is a sign that you do not have enough carbon in the mix. It should be damp.

With a reasonable mix of material, air and water a compost heap soon gets hot – as high as 75°C (167°F). At this temperature all weed seeds are destroyed. This heat is the energy given off by the bacteria, fungi and nematodes munching through your waste. In a well-made heap the population density of bacteria, actinomycetes and fungi can reach 10 billion microbes for every gram of compost. It is a mind-soaringly unimaginable number. The heat is given off as they convert the carbon to carbon dioxide. In fact, the same energy that a green plant starts out with during photosynthesis – carbon dioxide, water, minerals and energy – is generated in your compost heap. Everything is returned.

Turning compost half-way through its cycle. At this stage it is useable as a mulch but a little claggy and unevenly decomposed for use as a potting compost or for seed sowing outside.

By the time it reaches the fourth bay it smells attractive and feels pleasant, although it's still a little coarse in texture. This is ideal for using as mulch. However, it will be turned again and becomes finer and positively sensuous to touch and smell. When sieved it becomes a crucial part of our potting mix along with leaf mould, loam and grit, providing not just nourishment but that infinitely complex ecosystem to the plants. This means that they are not only healthy but also adapt well when they are planted into the soil. For this reason I would never sterilise soil or compost as it kills off all that is good as well as a few weed seeds.

But most of our compost is spread as a thin – just 2.5cm (1in) deep – mulch onto the soil once or twice a year depending on what is being grown. That is enough to trigger the soil into action and to build the humus.

Weeds

There is a group of plants that grow lustily every year, whatever the weather, and however negligent I am of their care. They never fail, and never let me down. Almost certainly they are same ones that are thriving best in your garden too, because they are all weeds. However you arrive at it, weeds are there in every garden and take up an awful lot of a gardener's time and energy.

A weed is more than just a plant in the wrong place. It is a plant hero, an adaptor and survivor, coping with any weather and outperforming all those around it.

Some weeds are hostile to the notion of horticulture from the outside but there are some that were introduced as garden treasures. The most famous of these is Japanese knotweed, introduced in 1825, and now officially Britain's most problematic weed and illegal to plant. Some simply become weeds in a particular garden because they do so well. Anything that self-seeds runs the risk of that fate as do perennials that creep out sideways, like *Lysimachia ciliata* 'Firecracker' in our Jewel Garden, or the royal fern that is swamping my Damp Garden, both of which started out as carefully nurtured treasures.

Yet before the gardener arms themselves with all the weapons of weed warfare and goes over the top into full-scale action against them, it is worth looking at them with a less belligerent eye. Many are beautiful. Dandelions, buttercups, ground elder and daisies all have lovely flowers, especially en masse (which, alas, they so often are). All are the ideal plant for the soil and situation – if not for our carefully constructed scheme of things.

The point about this is that weeds are an excellent indicator as to what will thrive in different parts of the garden. So, for example, docks indicate a deep, rich but rather heavy soil. Horsetail will only grow where there is lots of moisture, indicating poor drainage. Nettles are a sign of high levels of phosphate and nitrogen and always grow in damp areas where animals and humans have congregated.

Some weeds are now cultivated for their edibility. Many people eat young nettles in spring as a vegetable (like spinach) or as a soup, and I personally find them delicious – and they are exceptionally rich in iron. Ground elder, chickweed and fat hen have all been gathered and even cultivated as vegetables in the past. Who knows what prize veg will be rampant weeds in a few hundred years time?

All weeds are a nuisance but some are much more of a nuisance than others. Identify the real problems and put most of your energy into these. The real horrors are the perennials, and some combination of these is found in every garden in the land.

Perennial horrors

- ground elder
- couch grass
- creeping buttercup
- bindweed
- lesser celandine
- horsetail

Less disastrous

(but needing careful control if they are not to become a real headache)

- docks
- stinging nettles
- thistles
- brambles
- dandelions

There are a number of ways of dealing with these but the one thing that I cannot advocate is chemical control. I am not your man for that. But whatever ideology you may have about chemical use, my experience is that most weeds are controllable without them.

It goes without saying that prevention is better than cure. Always check plants that you buy or are given, especially woody plants that can easily host the roots of ground elder, bindweed or couch grass.

The second thing is to avoid the situation getting bad – or at least any worse. Tackle weeds as and when you notice them. In practice, this means that it is a constant job. But – and I think that this is really important and underrated – weeding is at the heart of gardening. I use it as a chance to get close to my plants and to judge the state of the soil, as well as part of keeping the place looking beautiful. So don't see weeding as a terrible burden imposed upon you but enjoy it as part of real gardening.

With all perennial weeds the best way to remove them is to dig up their roots. With nettles or brambles this is actually quite easy. But bindweed, ground elder and couch grass will regrow from the tiniest scrap of roots and as all are very brittle it is easy to break off a piece and leave it in the ground. Do not be discouraged. All of us miss bits of root. Just go back if you notice regrowth and take some more out. In time, you will get on top of it as well as dramatically weaken it and reduce the spread.

If this is not possible, either because the roots are entwined amongst a hedge or shrubs or because digging is just not an option, mulching is always a viable tool in the anti-weed armoury. It will not eradicate perennials but it will certainly weaken them. This is important, as if allowed to get healthy and strong the spread of some weeds is phenomenal. Bindweed can cover 25 square metres in one season and one creeping buttercup plant will colonise 4 square metres in a year.

The most effective mulches cut out all light and water from the weeds, starving their growth. A layer of 400–600-gauge black polythene will certainly do this. However, as well as looking horrible, it also destroys all other growth and is only useful for clearing ground prior to planting. A woven plastic 'landscape fabric' is a better long-term bet as it allows moisture through, but it doesn't look much better. When I use them I always cover them with a layer of chipped bark. This makes an effective path if the bark chippings are topped up each spring. I once trialled some mulch made from hemp but it was useless, as once wet it made a perfect growing medium for self-sown seedlings – invariably weeds – so made the situation worse.

On a border a loose mulch of organic material is needed. Anything will make a difference but for pure weed control heavy-duty bark chippings are excellent. Personally I find that they look rather municipal and prefer garden compost, mushroom compost, or 'strulch' which is sold in bales and is straw treated with the effluence from iron works to slow down its decomposition. I have used it for years and it is very effective – if not the prettiest of mulches. The advantage of these organic mulches is that they also feed the soil and improve its structure so whilst they inhibit weeds they also promote the growth of the plants that you value. But whatever you use must be thick to be effective. Really thick. A minimum layer of 5cm (2in) and ideally twice that or even more. It is always better to mulch a small area well than to try to spread it thinly over a bigger surface.

If all else fails it is worth constantly cutting back the growth of perennial weeds. This will weaken them and at the very least limit their spread. Often, especially for bracken and ground elder, this is the only feasible action and it does help.

So much for perennial weeds. I am afraid that there are lots of annual weeds to contend with, too. Their names read like a litany of amiable rogues: groundsel, chickweed, goosegrass (which is particularly bad in my garden), sowthistle, herb Robert, petty spurge and shepherd's purse – amongst many other offenders.

Dealing with these in the vegetable garden is simple. Hoe them and hoe them often. There is an old saying that if you need to hoe then you are not hoeing enough. Always hoe on a dry day, preferably in the morning so that the weeds will wilt and die in the sun (they can and will often regrow in the wet). The secret of hoeing is to keep the blade sharp and run it lightly under the surface of the soil in a push-pull action rather than jabbing at individual weeds.

I use a flame gun for our paths and this is very effective. It needs doing about once a month. However you have to be careful not to damage adjoining plants through heat radiation.

MAY

There can be no more jubilant moment in the gardener's calendar. April has a dancing green light that reveals itself between showers and moments of almost grumpy bad weather, but by the time we reach May there is a rolling, green richness that expands as the month progresses. Every year, May arrives like a gift and shakes me to the core, and it sends me spinning into a green space. No other time of year combines such an intensity of colour and freshness of light with the vivacity of daily – almost hourly – growth, and the full voluptuousness of the English garden on a perfect May day.

I find this slow green explosion almost beauty enough in both garden and countryside – and there is no moment in the year when I am more profoundly grateful to have British countryside on my doorstep.

Every day feels like a celebration – or even a thanksgiving – for having arrived, and every day throughout the month seems to build riches upon riches. At the beginning of the month the Jewel Garden ignites with tulips and wallflowers, and by Whitsun weekend at the other end of May, it is full with the green growth of all the herbaceous perennials, and sparked with colour from alliums, oriental poppies and the first irises.

Of course, no garden can compete with the British countryside in May where the combination of mile after mile of country lane and field hedgerow smothered with cow parsley and hawthorn blossom is, I believe, one of the great wonders of the world. It's an inspiration for any garden from inner-city backyard to a country garden like Longmeadow.

It is important to take the time to enjoy spring as it unfurls in the garden and world all around us. Time is a much-undervalued aspect of gardening and I am a great fan of Slow Gardening. Time flies fast enough without forcing it on. Slow

Gardening is subtle and considerate, gently steering and nurturing the garden rather than bullying it into some kind of chorus-line display.

But it also means hitting the regular rhythm of maintenance that makes such a difference to the way any garden looks, and which is the daily bread of any gardener. I mean jobs like weeding, dead-heading, clipping the hedges and mowing the lawn.

Finally, a very human but essential aspect that I love about gardening in May is getting rid of the layers that have protected me all winter. Gardening in a loose shirt and feeling the sun on my back is a reward for all those long, chill winter months.

ALLIUMS

Alliums have become a major part of our late spring display here. There was an element of accident in this. All our allium planting was done some 12 years ago in one week. Since then they have seeded themselves, spread and generally made themselves at home, and we have welcomed this eagerly because they are spectacular plants from the smallest chive flower to the crazy explosion of *Allium schubertii*.

The first to appear is *Allium hollandicum* (also known as *A. aflatunense*) in the Walled Garden. These are about 90cm (3ft) tall with lilac flowers fringed with a silvery halo. There is a white form, too. As they emerge, the colour showing through the thin tissue of sheath, they look like flat-topped thistles but then open out to a cylinder. The leaves, which fold half way up and droop in an idiosyncratic manner, have a tendency to start to die back at the tips before they come into flower. This is not due to any illness or problem – it's simply a common feature of early alliums that the foliage dies back before the flowers fully open.

Next is *A. h.* 'Purple Sensation' which has spread to the point of saturation in the Jewel Garden. It is truly both purple and sensational. The first one appears towards the end of April, although their foliage will start to appear as early as March, bursting free from its papery skin and by mid-May they are in their pomp. 'Purple Sensation' has a richness of colour that makes an ideal foil for the intense greens and yellows of May and June. As plants they require nothing of the gardener other than planting, and the best time to do that is autumn, although I have successfully planted the bulbs as late as February.

Allium cristophii follows in mid-May. They have huge exploding balls of flower, which work well with softer companion colours than 'Purple Sensation', which is why we grow them in the Walled Garden alongside old roses, sweet peas, geraniums

and other pale and pastel colours. As the flowers fade the dramatic flowerheads become drier and hold their shape for months on end, albeit without colour. We often collect them in this mummified state to put in a vase where they last for ages. Like 'Purple Sensation', they seed like mad and need thinning otherwise they can crowd a border.

On the other side of the Walled Garden we have *Allium schubertii*. This onion takes the exploding flowerhead to another level of artistic extreme. From a very short stem, the flower flies apart into a ball the size of a melon that is made up of dozens of flowers, each on a different length stalk. It can get lost at the back of a border so put it near the front. It dries easily and will stay in shape to be wondered at for months.

Allium giganteum is a colossal drumstick of a flower, reaching 2m (6ft) tall in our rich clay loam. It is almost like those box or holly bushes clipped into topiary pom poms and one year we had them marching all the way down the Long Walk. However I learnt to my cost (the huge bulbs are expensive) that it rots very quickly in a wet winter, and we lost the lot, so it needs as good drainage as possible. As a rule this applies to all alliums, which prefer a sunny, well-drained site.

Rather later – and last of the alliums here – is *A. sphaerocephalon*, which is much smaller than all the previous ones and in late July and August makes a tapestry of conical flowerheads of mauve and ruby pink. The only weakness they have on our heavy soil is that they tend to flop, but bounded by a box hedge they tumble onto its support rather than fall right over. Its natural habitat is a meadow and I have planted a batch of bulbs in the Writing Garden so next year I shall know if they, like the alliums in our borders, are happy to take up residence there and spread enthusiastically.

(Previous page) When this photo was taken in mid-May 2011 the replanted Jewel Garden was still relatively empty and dominated by the **Allium hollandicum** *'Purple Sensation' and the new foliage of the purple hazel,* **Corylus maxima** *'Purpurea'.*

AQUILEGIAS

Aquilegias, or granny's bonnets, don't just look pretty, but also carry an aura of wholesomeness about them. They are terribly easy to grow, unfussy about soil, seeding themselves everywhere and, being herbaceous perennials, come back year after year without any attention or horticultural care needed. They are happiest in the half-shade of woodland edge or shaded grassland, which in normal garden terms translates to the lee of a shrub or the middle of a border.

Direct sunlight and heat stunt its growth and singe the delicate leaves, which is a waste as they stay a good colour throughout the season, turning from a delicate glaucous doily to gain a violet tinge. The broad leaves have the habit of holding drops of rain after most other plants have slid the water away and the combination of the blue-green leaves tinged with violet, and the clear crystals of water, is enchanting. When the flowers are finished – around the end of June – cut the stalks off and let the leaves do their stuff without the distraction of drying spikes of stem.

There are about 70 different species of aquilegia and it is perfectly possible to use their colours with care and precision, and I have just planted out half a dozen healthy specimens of *Aquilegia vulgaris* var. *stellata* 'Ruby Port' into the Jewel Garden. These have, as the name suggests, a rich plum-coloured flower, multi-petalled, with yellow stamens. But they hybridise indiscriminately, producing bastard offspring with lots of hybrid vigour but flowers that are almost inevitably a dirty pink with more than a hint of green. The purist will despair, but I cannot dislike them, whatever their colour.

Although aquilegias are herbaceous perennials, in a mild winter the leaves linger to Christmas and new growth tentatively begins soon after, with seedlings popping up all over the place.

Astrantias hybridise readily and the result is often better than either parent. This one is growing through a box hedge in the Jewel Garden.

BIENNIALS

The great advantage biennials have over annuals in our borders is that they are hardy enough to withstand a cold winter and quickly produce flowers in spring, without having to wait for the plants to grow first.

One of the first spring biennials to appear in this garden is honesty, *Lunaria annua*. I planted a few individual plants about 10 years ago and they have self-seeded prolifically ever since. It germinates well from fresh seed, when the lovely lunar discs dissolve, and it quickly makes a clumpy rosette of leaves. These will overwinter and start to grow again in February. The seeds do not spread far, so they tend to mass in groups in a border and therefore need thinning every few years. The flowers are a characteristic mix of freshness and richness, often on the same stem, ranging from plum through to pale mauve. *Lunaria annua* 'Atrococcinea' is very red and 'Munstead Purple' is decidedly purple. The white form is less prolific for me but is mingled in amongst its purple kin with blindingly pure white flowers. Its pods are a pure green whereas the purple flowers produce pods with a purplish wash over them. Both become silvery when dry. They make a lovely cutflower both when fresh and when the seedpods have formed.

Wallflowers (*Erysimum cheiri*) are one of my favourite biennials – perfect for making a dramatic display mingled amongst tulips. It is a perfect combination making a rich tapestry of colour. Wallflowers have a distinctly honeyed fragrance intensified by sun and if you grow them in a confined area the scent can fill the whole space on a sunny day. We often grow them down the Long Walk and as you cross it, going down through the garden, it is like dipping into a fragrant pool.

Wallflowers are very easy to grow, although like all biennials you have to plan ahead. I sow the seeds in a seed tray or plugs in late May, pricking them out when big enough to handle, and I line them out in a corner to grow on until October when they can be planted in their final positions. They can be sown direct outside but will need thinning to at least 10cm (4in) apart if the plants are to develop strongly. They are members of the brassica family so are subject to clubroot, flea beetle and attack by cabbage white butterfly, so do not line them out where cabbages have grown the previous year.

You can buy young plants in September and October but make sure that they have some soil attached to the roots – just because they are tough does not mean that they are improved by neglect. They will grow in very poor soil although a bit of goodness gives the plants a leafy full body, which I like.

Forget-me-nots (*Myosotis*) need no growing regime of any kind. Indeed I know people that ban them from their gardens for being too invasive. Not me. I love and welcome them, but when they get dry they quickly get powdery mildew and at that point I pull them up – knowing that they will be back again next year however many I remove to the compost heap. There are a number of blue varieties to buy but 'Royal Blue' is taller than most. Like all biennials you can 'lose' a season of forget-me-nots by over-vigorous spring mulching. I transplant seedlings in winter by the spadeful, lifting and transplanting them like turf to break up over-large drifts and this spreads them around a bit.

Sweet rocket or dame's violet (*Hesperis matronalis*) arrives a little later, around the same time as the foxgloves, and the two look very good growing side by side. Sweet rocket has nothing to do with herb rocket and everything to do with adding a lilac pink leap into a border. *Hesperis matronalis* var. *albiflora* will give you white flowers, and the two come up side by side in our Walled Garden where it seeds itself in very erratic quantities from year to year. This irregularity is one of the charms of self-sown plants. If you want strict order and control (as with our wallflowers) then do it yourself, weed out all self-sown seedlings and start afresh every year. Go with nature and leave them untouched and you take pot-luck. I have seen it referred to as a short-lived perennial, but treat it as a biennial and you will not go wrong.

Foxgloves are a common native but as beautiful and exotic as anything growing anywhere in the world. This is **Digitalis purpurea** *'Pam's Choice' growing in the Walled Garden.*

Foxgloves (*Digitalis purpurea*) seed themselves right over the garden – especially in amongst hedges and shrubs – and we collect and move these seedlings although I stress that, like all biennials, this must be done in the autumn or late winter at the latest. Leave it to spring and they never really recover. The white form (*Digitalis purpurea* f. *albiflora*) will create beautiful ghostly plants for a shady border but subsequent seedlings often revert to purple, so it is best to sow fresh seed each year. We grow other colours such as 'Sutton's Apricot' and various yellow varieties and all are good and readily available.

HORSE CHESTNUT

I planted the horse chestnut, *Aesculus hippocastanum*, or conker tree, at the end of the Cricket Pitch (exactly where the stumps used to be) back in April 2002. I say planted but in truth I moved it there from The Mound because we were reshaping it with a digger, and the machine picked up the 2m (6ft) sapling with a good rootball and delivered it gently to its new home. It has grown happily there ever since and is now becoming a good-sized tree.

Horse chestnuts are at their best in early May, their great upright white or pink cones of flower by the hundred across each tree, then sticky buds break into leaf in early April and the whole tree is a delight all spring and summer. In hot days of high summer (let's hope) no tree casts such a congenial shade.

The trees were introduced into Britain in 1616 from their original home in Macedonia and Albania and quickly became popular as trees for large gardens and parks, though never really assimilated into British woodland. The timber of horse chestnut is soft and very white and was apparently used for dairymaids' buckets because, if kept constantly wet, it is slow to rot. However the demand for dairymaids' buckets is currently low. The pink-flowered form, *Aesculus × carnea*, is smaller so is often used as a street tree, but it never matches the beauty of the white-flowered original.

Iris sibirica *has a delicacy that few other irises can match. It likes heavy – even wet – soil and is perfectly at home in the Jewel Garden borders.*

IRISES

By mid-May the bearded irises in the Dry Garden are all flowering with rich, velvety purple petals but I confess that I do not know which variety they are. They were given to us, many years ago, by a friend who had also received them as a gift. None of us along the line could identify them. But a name would not make them any more beautiful.

Bearded irises come from areas with very hot summers and very cold winters. This makes them exceptionally hardy garden plants needing only good drainage and plenty of sun to thrive without any problems. They grow from rhizomes, looking very like generous ginger roots, which sit on top of the soil. When planting, it is important not to bury these rhizomes but to leave them so that they can become sunbaked. The roots that go down from the rhizome will gather all the moisture and nutrition needed to make the gorgeous flowers that come in every shade save red – hence the name Iris, from the word for rainbow and the female messenger of the Gods, whose emblem was a rainbow.

Through intensive breeding the flowers have doubled in size over the past 100 years. The price that we pay for this is that the bigger petals are more easily damaged – they are now much more susceptible to the ravages of wind and rain. To some degree, strength is being bred into certain varieties by increasing the thickness of the petals and the degree of their ruffles, but no iris enjoys wet, windy weather. Sun is their natural habitat.

There are thousands of named varieties and although many seem to be one particular colour, closer inspection shows that they are rarely pure. For me this is their glory, although I love the richness of the very dark varieties such as *Iris* 'Dusky Dancer' which is almost black, 'Ruby Chimes' which is a fabulously rich plum colour, and 'Jungle Shadows', 'Hot Spice', 'Provençal' and 'Wild Ginger' all of which surf through shades of brown. I have all these earmarked as additions to the garden.

Although you can plant them at any time, the best time to buy and plant irises is in midsummer, immediately after they have flowered. This is also the ideal time to divide the ones that you have already growing.

Bearded irises can become very congested, and as soon as they are a tangle of rhizome – which is likely to be every three or four years – they are best dug up after flowering and pulled or cut apart to make three or four new clumps, planted 15–30cm (6–12in) apart. It is best to throw away the oldest part of the rhizome because it will be much less vigorous than fresh, new sections. The leaves should be cut back to 15cm (6in), which is long enough to provide food for the new roots to develop but short enough to stop them catching the wind and rocking the plant before new roots have grown to stabilise them.

One of my favourite beardless irises that I grow in the Jewel Garden is *Iris sibirica.* This has small(ish) blue flowers tinged with mauve and yellow and orange tiger stripes at the base of the falls. It has a startling freshness and clarity. It will grow in the margins of a pond or quite happily in a mixed border, as long as the soil is not too light and contains plenty of organic matter. The foliage is much grassier than that of bearded irises and this can become a bit of a nuisance when it has finished flowering as, unsupported, these leaves flop all over the place. However, they apparently naturalise well in long grass, if the ground is rather heavy and damp, which seems to me to be a wonderful way to grow them – so I shall try an experimental clump or two in the Writing Garden for next year. *Iris sibirica* is available in colours other than blue, such as 'Ruffled Velvet' which is amethyst, and the claret-coloured 'Showdown'. But I love the clear blue the best.

Iris chrysographes grows in similar conditions to *Iris sibirica* and has breathtakingly rich, velvety flowers of the darkest purple and 'Black Knight', as its name dares to boast, is almost truly black. We have planted it at the edge of a path in the Jewel Garden where it sucks light into its inky depths.

ORIENTAL POPPIES

There is always a delicate, silky revelation, which tells me that spring has evolved into summer. This has nothing to do with the weather or calendar, although it usually takes place somewhere around Whitsun and the weather seems to be poised at that apex of benign lushness that makes Britain in May the most beautiful place in the world. But come rain, chill winds or burning sun my personal measure of this seasonal shift is that moment when the oriental poppies first come into flower.

I do not think that any other plant produces a flower that is so saturated in colour and yet miraculously manages to balance this both within itself and the garden burgeoning around it. The twirled cone of flower emerges from the downy green apricot of the bud and the petals leisurely unfurl from a ruckle of slinky silk. Botany is never more voluptuously sensuous. For three weeks the poppies dominate the perennial borders then they collapse in an increasingly crumpled heap, all colour exhausted like gorgeously begowned revelers at dawn.

It is always a race to see which opens first, the huge crimson flowers with a black blotch at the base of each petal of 'Beauty of Livermere' in the Jewel Garden or the smoky mauve petals of 'Patty's Plum' in the Walled Garden. Although very different, both are stars that take centre stage whilst in flower.

The flowers of the natural *Papaver orientale* are a bright orange and the range that we are now familiar with of pinks, oranges, whites, reds, and plum has only been available since 1906 when Amos Perry, the nurseryman, noticed a pink form amongst his red seedlings which he named 'Mrs Perry'. He then, after trials and tribulations, bred the first white oriental poppy, 'Perry's White'. If you want a more unquestionable monochrome then 'Black and White' is even more starkly delineated with its deep black central blotch.

Since then dozens of different colours and scores of varieties have been carefully raised covering every permutation of red, pink and orange. As well as 'Beauty of Livermere', we grow 'Ladybird' which has enormous vermilion flowers, often very early, and some of 'Pizzicato'. These are rather short – so good for the front of the border – and each plant will produce a range of colours from white to quite dark red.

We support the poppies before they come into flower because the heavy leaves and big fat flowers can flop dramatically. As soon as they finish flowering I then shear them back to the ground and remove all the existing leaves to the compost heap. This is dramatic but lets in light, air and water, and they then regrow to give a second flowering later in the summer. It also makes space that can be filled with a tender annual that will have enough time to get established before the poppy foliage begins to regrow.

Although they grow easily from seed, self-sown seedlings are usually a muddy colour – the worst of the parents – and it is better to lift the plants in autumn, divide them with a spade and replant the pieces. They will start growing immediately and overwinter with small leaves before taking off with fresh vigour in early spring. They also take easily from root cuttings. Cut straight sections of root into 8–10cm (3–4in) lengths, the thickness of a pencil. Insert these vertically into a gritty compost mix so the top is just below the surface. Water and place in a cool greenhouse or coldframe. They will be ready to pot up individually the following spring.

PURPLE FOLIAGE IN THE JEWEL GARDEN

Purple leaves will make red and yellow flowers seem more intense, and add more depth to a border than any green can, as well as being attractive in their own right, so they are particularly suitable for the Jewel Garden.

I was first struck by how effective purple foliage can be in the garden of Kiftsgate Court near Chipping Camden in Gloucestershire. The way that purple hazels were coppiced there influenced me dramatically. Coppicing the shrubs – in other words regularly cutting them right to the ground – produces vigorous new shoots that carry extra large leaves. After that visit to Kiftsgate I bought some purple hazels, *Corylus maxima* 'Purpurea', and sat back to watch them become healthy plants.

Our coppice is made up of 73 green hazels, *Corylus avellana*, which all grow lustily, so I assumed that their purple cousins would perform just as well in the Jewel Garden. No such luck. For years they hardly grew at all and clung on to life in such a shrivelled form that I eventually dug them up and replanted them in a nursery bed, where, to my surprise, they flourished – not only producing purple leaves but also pink- and magenta-tinged nuts. Finally I worked out what the problem was.

Although planted in an open border, they were nevertheless being shaded out by the vigour of the annuals that I grow in those beds, particularly *Atriplex hortensis* var. *rubra*, the purple orach, which is a prodigious self-seeder and which we love for the purple foil it provides to all the surrounding rich jewel colours. It took me a while to realise that an annual with a life of just a few months could crush the will to live in a small tree. In the end I bought four more much larger purple hazels and planted them in exactly the same spot and they thrived – because their foliage was never competing for precious light.

However, I do not coppice right to the ground because the young shoots would face exactly the same problem as the first batch of plants, so I selectively prune a third down to the ground every winter, leaving enough up in the light to nourish the roots. It works well.

Lysimachia ciliata *'Firecracker' has tiny bright yellow flowers in summer but we grow it for its chocolate leaves which are a good foil for neighbouring plants – although it can become invasive.*

If they had green leaves they would probably have coped with the lack of light but the red layer of pigment over the green of the chlorophyll means that there is less photosynthesis and therefore less food. In short, purple-leaved plants need more pampering than green-leaved ones and, if they are to grow at all strongly, more light. Yellow-leaved shrubs are also weaker than green ones but are liable to scorch in direct sunlight.

I have in the past thrown away an *Acer palmatum* var. *dissectum* Dissectum Atropurpureum Group even though I wanted what it had to offer and it was alive, because it wasn't growing properly and I assumed that my soil was not acidic enough for it. That was nonsense. They will grow just about anywhere that is reasonably sheltered. It was just too shaded.

As well as the hazels in the Jewel Garden borders we also have a few other purple-leaved shrubs to act as a backdrop to increase the intensity of colour – especially in late summer. There is the smoke bush, *Cotinus coggygria* 'Royal Purple', which has excellent burgundy leaves with – like the wine – just a touch of brown. Although I am not sure if I like the smoke (ie the flowers) from the smoke bush – it is a bit like coloured smoke from a disappointing firework. The leaves of *C. c.* Rubrifolius Group (syn. 'Foliis Purpureis') are too matt and brown for my taste, but 'Velvet Cloak' is good. I also have a purple elder, *Sambucus nigra* f. *porphyrophylla* 'Guincho Purple'. This is so vigorous that it will take a really hard prune every year, outgrowing the competition for light around it, but the cotinus needs plenty of light, especially in early summer.

You do not have to use shrubs for your foliage plants. I have mentioned the purple orach that we grow, which can become almost a weed, although the seeds need light to germinate so a thick mulch in early spring snuffs them out. I also grow *Lysimachia ciliata* 'Firecracker' which is a perennial with chocolate leaves and little yellow flowers. It is a superb foil if you have dampish soil – although it flops badly in dry conditions. However, if happy, it spreads like wildfire and needs reducing by as much as three quarters every few years to keep it under control.

The intense foliage of the four purple hazels, **Corylus maxima** *'Purpurea' tighten the intensity of colours around them. This new growth will be 3m (6ft) tall by the end of the summer.*

ROSA CANTABRIGIENSIS

At this time of year all the roses are starting to come into flower with more or less enthusiasm depending on variety and situation, but there is one rose in the Spring Garden that is reliably at its best at this time of year and that is the species rose, *Rosa cantabrigiensis.*

It is the first to bloom in this garden, its little snouts of primrose flowers breaking along the length of its branches against the fern-like leaves. These then open into discs of fragrant, gentle yellow flower that last for about three weeks, just as most of the other, more familiar garden roses are getting into their stride.

Species roses are invariably tough shrubs that need no special treatment and certainly no pruning or feeding once they are established. But for all their hardiness, I believe that they are some of the most beautiful of the rose family.

R. cantabrigiensis was discovered in Cambridge Botanic Garden and is a cross between *R. hugonis* and *R. sericea*, both of which I have growing here at Longmeadow. It has chocolate stems and ferny, delicate foliage of a beautifully apple-fresh green. I have two bushes in the Spring Garden and they both have a tendency to sprawl, but not in an ungainly fashion. It is best to plant them where they have room to fully express themselves, although one of mine is near a path and has to be regularly cut back. Left unpruned they will reach about 2.25m (7ft) high and be about 1.5m (5ft) across. They will grow in almost any soil and in full sun or part shade.

SPECIES ROSES

Species roses (or species anything) are those that occur naturally without variation. They will come true from seed when pollinated by themselves or others of the same species. Obviously, if you cross two species then you get a hybrid, but the original species parentage will always be there to a degree, however tortuous the breeding line becomes.

So with species roses there is no clumsy trace of the breeder's hand. If they tangle and twist then that is what they do. If they only flower for a few days then that is unchangeable. They are purely themselves without any additions or dilutions. This imbues them with a freshness and wildness that I love.

These roses are a celebration of the lightness and freshness of spring and early summer and their effect in a garden is as a result of the overall combination of flowers, foliage, stems and thorns, rather than the intensity of a single gorgeous bloom. The sum of any individual species rose's parts tends to amount to more than the whole. That is their great charm.

One of their great attractions for the nervous gardener is that species roses are amongst the toughest plants in the garden and will grow in almost any soil and any position. They rarely suffer from the afflictions of their more highly bred cousins.

The first to appear in my garden is *Rosa sericea pteracantha*, which I grow primarily for its famous shark-fin thorns, but the flowers, for such a barbed plant, are surprisingly pretty. *R. sericea* is the only rose with only four petals on each flower. The thing to note when growing *R. sericea* (or any of its variations, such as the fabulous 'Red Wing') is that the best thorns are produced on new shoots but the flowers are on old ones. I get round this by pruning it every other year rather like a dogwood, removing a third of the oldest stems right down to the ground.

Around the time that the native species roses such as the dog rose (*R. canina*) and the sweet briar (*R. eglanteria*) appear in the hedgerows, *R. cantabrigiensis*, *R. primula* and *R. hugonis* start to flower in my garden, all with ferny foliage and lovely primrose-yellow flowers. One of these would probably do but it has never felt like too much yet. I cannot recommend these roses highly enough, and they exactly fit the mood of the Spring Garden in May, which is fulsome but light and frothy before the canopy of the trees closes it down for the summer.

Species roses will take you right through the season into autumn. I have a *R. moschata* growing in the Coppice that only begins to flower in the second half of September and is at its best in the second week of October. They can be ground covering, mounding or rampant rompers reaching 15m (50ft) up into a tree.

Many of the species flowers turn into lovely, curvy hips. *R. rugosa* are like tomatoes, *R. moyesii*, growing in the Grass Borders, are like miniature orange bottle gourds (although the dry, cold spring and summer meant that 2011 was a very bad year for these hips). Those of *R. pimpinellifolia* are a deep brown and those on *R. glauca* are bunched like grapes. *R.* 'Wintoniensis', in the Walled Garden, has great clusters of flagon-shaped hips that have a curious purplish bloom.

If you want to make the most of these hips then you should not prune until late winter – and then only to remove dead or damaged wood or to restrict the size. But if you want to make the most of the flowers the time to prune is immediately after flowering. But don't fetishize it. Just cut out dead wood, take off any stems spilling across a path and let the shrub get on with it. There is nothing that man can do to these plants to improve them.

TENDER ANNUALS

We are always boosting the late summer colour in the Jewel Garden with half-hardy annuals. These will not last beyond the first frosts but add excellent colour, structure and a touch of glamour to the garden well into autumn.

They are often not easy to buy as plants from a garden centre but the seeds are widely available and can be raised in plugs and pots under cover or sown direct once the soil has warmed up. I sow them between mid-April and mid-May so that we have a staggered supply ready to go into the garden from early July through to late August. All these plants respond to heat rather than light, so if we have a mild autumn they can and will flower right into November.

The first, and one of my favourites, is *Tithonia rotundifolia*. It has bright orange flowers with a yellow centre, drawn with child-like simplicity, sitting on a long, slightly funneled hollow stem that has an extraordinarily velvety texture.

It can grow to 1.5m (5ft) and will go on flowering until frost destroys it. 'Torch' is a very good variety but a little smaller – perhaps 1.2m (4ft) – and 'Goldfinger' smaller yet, keeping to 90cm (3ft) tall. 'Aztec Gold' is tinged with yellow. They like maximum sun and a rich, well-drained soil. They need staking as the leaves are large and catch the wind.

Cosmos bipinnatus also has simple, daisy-like flowers but the foliage is finely cut and loose and they mix well into a border without dominating it. I love the white 'Purity' and the slightly smaller 'Sonata White'. 'Dazzler' is a rich magenta with a gold centre.

Leonotis leonurus is an extraordinary plant. It will reach 3m (10ft) tall with bright orange flowers sprouting all around the ribbed stem. But in a cold summer, like 2011, it is very reluctant to grow much at all. It is actually a tender perennial but will not survive a hint of frost so is best sown fresh each spring.

Finally zinnias make one of the best cutflowers. They are prone to slug and snail attack so watch them carefully and mulch with grit. *Zinnia elegans* 'Envy' is famously green-petalled but the mixed varieties such as 'Early Wonder' or 'Parasol Mixed' have an excellent range of bright, even garish colours. Giant cactus zinnias look like mini dahlias – but are larger than the mixed varieties – and the scabious-flowered varieties are also large and have excellent colours.

Zinnias are tremendously good value in the Jewel Garden radiating intense colour for months.

We always grow **Cosmos bipinnatus** *'Purity' for the Walled Garden where it flowers well into autumn.*

UMBELLIFERS

Nothing celebrates the glorious month of May better than the wonderful froth of cow parsley running along a thousand miles of British country lanes, often beneath the equally lovely display of white hawthorn (may) blossom.

Cow parsley is, I suppose, a weed, although few garden plants ever look as good to my eye. But it can certainly spread rampantly in a border given the chance and I have it in my Spring Garden as an honoured guest. Less invasive is *Anthriscus sylvestris* 'Ravenswing', a cultivated variety with purple leaves and brown stems beneath the white, lacy flowers. It cross-pollinates with the wild cow parsley, which means that the offspring quickly lose the intensity of purple leaves, so if its purpleness is the main attraction for you then keep it well away from its wild cousins. I confess I do not mind this hybridisation as I am besotted by cow parsley in all its variations.

The flowers are carried as upturned umbels – hence the description umbellifer. All umbellifers are a good thing in the garden because they attract a range of beneficial insects. In fact, probably the best action that the gardener can take against aphids is to encourage ladybirds, hoverflies and lacewings into the garden by planting plenty of umbellifers like dill and fennel, or letting a patch of carrot go to seed. Birds, too, are attracted in autumn by the seeds, so they instantly enrich the wildlife of your garden.

Many umbellifers add a tall, even stately element to the garden whether growing in a border or wild at the fringes. The biggest umbellifer of them all, bigger even than giant hogweed, is *Ferula communis*, the giant fennel. We have this planted in the Grass Borders and in time it will develop a flower stem 3.7m (12ft) tall, more like an agave flower than British woodland froth. As it is monocarpic, this great performance is also the plant's death throe. If you buy a plant from a garden centre be patient, as

it can take a few years to form a flower stem and set seed, although when it does so it will grow at an astonishing speed so by late spring it is up to 3m (10ft) tall. But even without the stem, the exceptionally finely cut leaves are worth their place.

Angelica is huge – and, unlike giant hogweed, perfectly legal – but can become a weed, too, thanks to its great scatter of seed which, given dampish, rich soil, will become a thicket of stately plants. Nothing wrong with any of that, and we treasure it, although ruthless thinning is needed. It is monocarpic so the parent dies once the seeds are shed. *Angelica gigas* is smaller, a deep plum colour and perhaps more adaptable for a small border, but certainly less intrusive. After a modest early summer it comes into its own in August (see also page 210).

The common fennel (*Foeniculum vulgare*) is one of the welcome weeds in my garden, spreading liberally all over the place. It originates from baked Mediterranean hillsides where it survives as a surprisingly scrawny plant with all its intense aniseedy flavour loaded into a condensed frame, but grown on our rich soil it will reach 2m (6ft) tall. We let the bronze form seed itself freely throughout the garden, but particularly in the Grass Borders, as much for the stems – that can be like the best bamboo – as the umbels of flower, although these have their finest hour when they twinkle with a November diamante frost. Fennel seems to exemplify all the umbellifer virtues: open, lacy, towering but not shading anything beneath or around it. The leaves are the perfect accompaniment to a baked or barbecued fish and the seeds are delicious too, both rubbed into a joint of pork and to munch on by the handful. Good for the troubled tummy, too.

Two other medicinal and culinary umbellifers that we grow mainly for their looks are chervil and sweet cicely. Chervil, *Anthriscus cerefolium*, grows best in light shade and will go to seed very fast in full sun. The idea is to gather the leaves and eat them before the plant flowers, so to guarantee a fresh supply of flowerless plants you need to sow seed monthly. Sweet cicely, *Myrrhis odorata*, looks like a more feathery version of chervil, but it is not related. We have it growing in quite deep shade in the Coppice, where it seems very happy. It appears before and stays longer than other umbellifers and has the freshest green leaves of any plant. All of it can be eaten, from root to flower and seed, and all tastes of aniseed. It is good cooked with tart fruits such as rhubarb or gooseberries to reduce the acidity, and when added to cream it cuts the fattiness.

TOMATOES

Tomatoes are the most popular of all edible garden plants, so clearly a lot of people know how to grow them perfectly well. But this is how I do it and it works well enough for me, given that my priorities are taste first, healthy plants and fruit second and the size of the crop third.

I sow my seed around the end of February. These are all destined to spend their lives in a greenhouse. If you are growing plants to be grown outside there is no point in sowing before early April as they will not grow well until the nights warm up in June. Tomatoes need some heat – around 15°C (59°F) – to germinate, so they need a heated greenhouse, a warm windowsill or a heated propagating mat.

As soon as you can see two true leaves with zigzag edges, carefully pot them on into peat-free potting compost in large plugs or individual 8cm (3in) pots. Keep the growing seedlings warm and well-watered, but they will grow fairly slowly until light levels increase in April. If they are cold young tomato plants will develop blue leaves but will quickly grow away strongly when warm. I plant my greenhouse tomatoes in mid- to late May according to the weather and wait until June before planting outdoor varieties (having first hardened them off well).

There are two types of growth, bush and cordon, which you may see referred to as determinate and indeterminate. Bush tomatoes grow to a vigorous bushy size, the fruit all ripens more or less at once and are harvested. Cordon tomatoes will keep growing for as long as the conditions are right, reaching astonishing heights and producing fruit over a long period, although our climate restricts this in most places to early July until late October.

Cordon tomatoes are trained up a cane or twine and are pruned by pinching out all the lateral shoots that grow at 45 degrees between the stem and leaves. These shoots are extremely vigorous and take too much energy into plant rather than fruit growth. Bush tomatoes need no pruning at all but take much more space.

Young tomato plants growing in the top greenhouse with a robust support system rigged up in place ready to bear the load of a heavy crop.

I plant out cordons in beds in the greenhouse with each plant about 45cm (18in) apart and 60cm (2ft) between the rows. This is quite close but as long as the greenhouse can be well ventilated they grow healthily enough. Although they are close together there is plenty of room between and around the rows for picking, pinching out and for air to circulate.

I first make a strong bamboo support structure like mini scaffolding to tie the plants into as they grow. This means that the plants never break under the weight of fruit, which by August can be considerable.

Always plant tomatoes deeply so that the stem is buried right up to the first pair of leaves. New roots will grow from the submerged part of the stem that will both anchor it more securely and provide more food and water for the mature plant.

Allow at least 90cm (3ft) in each direction between bush plants and preferably twice that between rows. I support bush varieties with a couple of canes and string. The basic principle is to simply stop them flopping over.

If you are growing in containers, a 20-litre pot is good and a 15-litre one is about the smallest. Peat-free growing bags are surprisingly successful but can be much improved by extending the root depth using a couple of old pots with the bottoms cut out. Fix the pots into the bag, then plant the tomatoes into the pots so the roots grow on down into the bag beneath them.

Tomatoes need plenty of water, especially while they are growing. I water every three or four days in the soil and every day in pots. As the fruits start to ripen this watering regime can be cut back, otherwise the skins may split.

If you have good soil, well enriched with compost, the plants will not need feeding. But container-grown plants do well with a feed once a week. I use liquid seaweed or homemade comfrey fertiliser, and lay any spare comfrey leaves on the soil around each plant, which not only feeds them but also helps retain moisture.

Tomato problems

Most problems can be avoided by maintaining a steady heat along with good ventilation. To help the airflow (which is the best defence against fungal problems) it is a good idea to remove the lower leaves up to the first truss of fruit as they set. When these have ripened the next set of leaves can be taken off, and so on, until by the end of August the plants are completely leafless. There is enough photosynthesis via the stems to keep the plants growing healthily whilst the remaining green fruits ripen.

Tomato leaf mould This manifests itself by a yellowing of the upper leaves and a greyish brown mould on the underside, which quickly leads to defoliation. At the first signs of this all affected leaves should be removed and burnt. This soil-borne mould is often a residue of previous crops like lettuce and is encouraged by poor ventilation and too much humidity.

Blight Tomatoes are close cousins to potatoes and suffer the same diseases, especially potato blight. The blight manifests itself as pale brown blotches on the leaves that will quickly radiate out. The fungus can get to the fruit making them inedible. At first signs remove all affected leaves and spray with Bordeaux mixture (a combination of copper sulphate and lime).

Blossom end rot shows as a flattened, calloused, hard brown disc at the end of the fruit. It is caused by inadequate water supply, which in turn stops the plant from taking in enough calcium so that the cells collapse. Water regularly and, if your soil is very acidic, grow only small-fruited varieties that tend to be less susceptible.

Splitting fruits and leaf curl Fruits sometimes split and develop a grey, wispy mould. It is caused by an irregular water supply or big temperature variations from day to night. Close the greenhouse before the evening cools down and keep it wide open all day. Cold nights will also cause the leaves to curl up lengthways and look as though they are about to die. Older leaves are more affected than young ones.

White fly The adult white fly lays its eggs on the lower leaves, and after a nymph stage the new adult emerges and feeds on the leaves, sucking sap, spreading viruses and exuding honeydew on which a fungus grows. The flies overwinter, especially on perennial plants, so try not to keep plants like fuchsias overwintering in a greenhouse that is to grow tomatoes the following summer. I grow basil with my tomatoes as an effective deterrent and the basil thrives in the tomato-growing regime.

Varieties

- 'Shirley' (Unspectacular but nice, and a banker as very disease-resistant)
- 'Andine Cornue' (Wonderful horn-shaped fruits. Great for cooking)
- 'Black Russian' (Looks odd but tastes superb)
- 'Brandywine' (Huge beefsteak fruits. Great sliced raw and for sauces. Slow to grow but worth the wait)
- 'Tigarella' (Very tasty striped fruits. Good raw and cooked)
- 'Gardener's Delight' (Another banker but tasty and very good roasted on the vine)
- 'Costoluto Fiorentino' (Large, ribbed meaty tomato. Rich taste)
- 'Principe Borghese' (Small, almost rectangular, lovely raw and cooked)
- 'San Marzano' (A plum variety, considered by Italians to be the best for tomato sauce)
- 'Marmande' (Big plant, big fruits, big taste. Will ripen outside in most summers if in a sunny spot)

'Black Russian' tomatoes ripening. Their colour can vary from a muddy cast of green over orange to an almost chocolate stain.

ASPARAGUS

I made a new asparagus bed this year. This is not something to do lightly. The last time I planted asparagus here was in 1999. In fact, once a decade is almost the stuff of makeovers when it comes to asparagus. A good bed should last a generation. It is a vegetable for the long haul.

You can of course buy fresh asparagus pretty easily. Evesham has long been the centre of asparagus production and nowadays it is grown by the mile under plastic in Herefordshire. This is certainly better than no asparagus at all (heaven forfend!) but really nothing like as good as spears cut from your garden and rushed straight to the boiling pan of water. Like new potatoes, peas or sweetcorn, asparagus is one of those harvests that diminishes in quality almost by the minute.

You can grow asparagus from seed although it is more usual to buy crowns or one-year-old plants. They are best planted about 30cm (12in) apart and with 15cm (6in) of soil above them. It is important to spread the roots very carefully as they are brittle and easily broken but do not, under any circumstances, let them dry out. I left a batch in the sun for just half an hour once and not one grew as a result.

If you start with healthy crowns planted when the soil has warmed up in spring you will take your first – modest – harvest the following year, a more substantial but restrained one the following year, and tuck in with relish in year three. Each established plant can produce 30 or 40 spears over the season, which in my garden runs from the end of April to Midsummer's Day. In warm, humid weather the spears can grow 23cm (9in) in 24 hours whereas in a cool week in April or early May there can be very little growth for days at a time – although a layer of fleece or cloches in early April will help produce an extra early cutting.

Traditionally it does best in sandy soil though it grows well on any fertile ground, but if your soil is on the heavy side it needs to be grown on a ridge to keep the plants out of wet soil. Drainage is the key, however that is achieved. The choice is either to plant on a double mound, one for the roots as a ridge in a trench below the surface, with the extra soil mounded up above that which the spears will then grow up through, or simply to add lots of grit to your soil and plant the crowns 15–23cm (6–9in) deep straight into the ground. My first asparagus bed at Longmeadow was of the ridge-in-a-trench variety but this new one in the Ornamental Vegetable Garden is simply one of the beds with added horticultural grit.

Asparagus can be affected by fusarium wilt, which will manifest itself with red streaks at the base of shoots before the inner root tissue collapses. On top of wilt you might get asparagus beetle, asparagus fly, asparagus rust, purple spot and violet root spot, as well as the slugs that like to nibble the emerging spears. But, in practice, any of this beyond slug damage amounts to bad luck or bad drainage. Male plants are likely to be healthier than female ones. Female plants produce berries and seed from around late August whereas male plants have more vigorous fern, which in turn feeds the roots better and makes for a better crop the following spring. 'Lucullus', 'Grolim', 'Purple Passion', 'Backlim' and 'Gijnlim' are amongst quite a few all-male varieties that are now available.

You should stop cutting spears in July and let the shoots develop ferny foliage. I leave them to extract maximum goodness from the sun and transfer it to their roots, cutting them back in autumn when they turn yellow. Then mulch them in early spring with good garden compost and do nothing else other than keep the bed weeded.

BEETROOT

I remember filming in Poland 20 years ago and coming down to breakfast to be greeted by a whole beetroot, bread and yoghourt. It was something of a culture shock. However, eastern European countries have always celebrated the virtues of beetroot whereas although most gardeners will grow some, I suspect it is eaten with less enthusiasm than many other vegetables. This is a pity, because eaten hot – either roasted with some thyme or boiled and then served with a hot cream sauce – it can be truly delicious. But perhaps not for breakfast. The secret is to grow two or three batches a year and to eat them when they are young and tender.

Beetroot will not germinate at temperatures below 7°C (45°F) so I sow two or three seed clusters per plug (not too small) and germinate them in the greenhouse before hardening them off in a coldframe. There was a time when I always thinned beetroot conscientiously but I now like to grow them as a little group of between two and five beets. I prepare their growing ground with a good dressing of garden compost. It is a big mistake to treat them like carrots or parsnips and starve them. They respond well to plenty of richness and water and if too dry or unnourished will respond by bolting which will stop root development unless the flowering stem is cut back as soon as it is noticed. When the seedlings have been properly hardened off – and the simplest way of doing that is to put the plug trays on the soil where they are

PLANTING BEETROOT SEEDLINGS

Thinning seedlings in a tray of plugs to leave one to three strong seedlings for each individual plug.

When the seedlings have a decent root system that holds the soil of the plug together when lifted they are ready to be planted out into the garden.

I space them 23cm (9in) apart in rows divided by the width of a board I kneel on to avoid compacting the soil. Straight lines make hoeing a lot easier.

Finally give them a really good soak after planting and make sure that they do not dry out too much as they grow.

to be grown for a week – I plant them out at 23cm (9in) spacings in each direction. This is wide enough to get a small hoe in between them and also for the groups to swell out with ease. I make three or four sowings between March and August to keep a continuous supply. The final sowing will overwinter well into spring.

There are many different varieties to choose from in a whole range of colours and shapes, but I particularly like 'Chioggia', 'Burpee's Golden', 'Pronto' and 'Carillion'. They are a good crop to grow in a container as long as there is at least 23cm (9in) depth for the roots.

CHICORY

Chicory adds a deliciously bitter tone to a salad and is not something that is likely to appeal on its own, but it will transform an otherwise bland mix of leaves. I love them and love growing them.

All chicories are easy to grow and do so in two stages. The first, from mid-spring to late summer, develops the root system, usually with a mass of green leaves, which are perfectly edible, especially cooked as they are very bitter at this stage. However, after about three months these first leaves undergo a change. This transformation differs from variety to variety, with some dying right back, some changing shape (like 'Grumolo Verde' which starts out as a loose-leaf and then slowly develops round, almost spiralled heads) and many changing colour from vivid green to bright red in response to daylight length and temperature. Some varieties have a kind of self-blanching mechanism where the outer leaves protect the inner ones from light which results in these being paler and less bitter, and the 'witloof' chicories are best-known for responding to artificial blanching either indoors in the dark or under a forcing pot.

I have grown over a dozen types of chicory but there are two that I return to. The first is radicchio – the cabbage-shaped red balls of leaf, and the second is endive. The two radicchio varieties I like best are 'Rossa di Treviso' and 'Rossa di Verona'. Sow them in May or early June, thinning them to 23cm (9in) spacing (this is important – they need plenty of room). Keep them weeded and let the loose green leaves grow. In autumn these will start to droop and should be gradually removed from the outside. They make excellent compost. Inside the heart of the plant new red foliage will appear to form the familiar tight crimson hearts. Remove any green leaves

falling over the heart. Chicory is very resistant to cold but hates damp air so I cloche them in very wet winter weather, removing the cloches when it is dry. Cut the hearts when any size from tennis ball upwards and they should regrow right into next spring.

Endive is the second chicory that I eat as 'Cornet de Bordeaux' or a hardier curly-leaved one like 'Coquette' or 'Frisee de Ruffec' but all will grow exactly like a lettuce and add depth to any spring or summer salad. If you find them a little too bitter then try blanching them by tying up the leaves with string for a couple of weeks before cutting. Unlike radicchio, they are susceptible to bolting in hot weather so sow fresh supplies every few weeks.

GOOSEBERRIES

Raw, stewed, in a pie, crumble, fool or ice cream – however you choose to eat them, gooseberries are one of the least appreciated and best of all fruits. And of all fruits they are the least trouble. Indeed, a gooseberry thrives on benign neglect.

I learnt this the hard way. When my batch of assorted gooseberry bushes – 'Invicta', 'Langley Gage' and 'Whitesmith' with a couple of the red-fruited 'Whinham's Industry' – were first planted I treated them with reverence and respect. Their planting holes were lavished with manure and compost, they were sheltered from the nasty bit of wind that tears through that corner and each March, after I had carefully pruned them, I mulched them with liberal amounts of compost. In short, they were loved.

All this care was repaid with mildew and sawfly on a pestilent scale. They grew vigorously and produced lots of fruit but not a single bush escaped fungal or insect attack. This went on for a few years whilst I responded by lavishing even more care on them – which was the worst thing possible. Then I met an old lady who said that her father's gooseberries were his pride and joy. Not only did they eat enormous gooseberry pies every year but also he regularly won first prize for them at the local show. His secret was neglect. The bushes were in a shady corner near the back door. Once a year her mother emptied a bucket of wood ash over them and her father pruned them with shears, but that was the sum of all their horticultural care.

Initially the leaves of 'Rossa di Treviso' are green until, as here, they start to turn red in late summer, and edible new leaves emerging in autumn are deep crimson.

So I moved my ailing bushes to a new site exposed to a wind that howls across the fields. I added nothing to the soil they were planted in and gave them a mulch of ash from the fire in spring.

Things improved from the first year. Sawfly – which had rampantly defoliated them with monotonous regularity – practically disappeared. This is because sawflies hate wind. The adult fly lays its eggs at the base of the gooseberry bush and the caterpillars slowly eat their way to the top, which means that you often only notice the damage when the foliage has been almost completely eaten. But on a windy site the adults will move to a sheltered spot to lay their eggs.

When they did have leaves they were often covered with a grey mould, especially later in the summer. This was American mildew, *Sphaerotheca mors-uvae*, which thrives on moist, stagnant air. There is little one can do about the moisture but wind helps more than anything to keep it at bay.

I also pruned all my bushes to have a central stem or 'leg' about 30cm (12in) high, which allows the air to get under the bush, and now I prune all inward growth every year. This is radical pruning but maximises light and air at the centre. Having said this, the soft new growth that pruning stimulates is what rampant fungi and hungry caterpillars love most. An old, unpruned gooseberry bush rarely gets troubled by sawfly or mould presumably because it is too tough a mouthful. But if you love gooseberries – and I do – then they are so much easier to pick from an open, goblet-pruned bush or a cordon.

KITCHEN HERBS

We are hearing a great deal nowadays about the resurgence of vegetable growing – and three cheers for that – but do not overlook growing herbs, too. A small patch can provide masses of herbs that will be cheap, easy and improve the simplest dish. There are two prerequisites for kitchen herbs. The first is that they must be sufficiently abundant to use them unselfconsciously so you stop thinking of them as a garnish but use them really generously. The second is that they must be convenient for collecting. Key kitchen herbs must be close enough to leave a pan on the stove whilst you nip out

There are many varieties of basil but all taste best if grown quickly with plenty of heat and moisture.

BASIL
ITALIANO CLASSICO

and grab the required handful without rushing and without burning anything. My own culinary herb list is bay, rosemary, sage, thyme, French tarragon, oregano, fennel, chives, mint, parsley, basil, garlic, coriander and dill. There are others of course, but this is a really good selection.

To start with the biggest – and the biggest loss. Our lovely 15-year-old bay tree was half-killed a few years ago but grew back only to be frozen to death the following winter. I have planted another in a large pot and will bring it indoors if we have another really harsh winter. However, given mild weather bay will become a beautiful and large tree with its marvellously aromatic, evergreen leaves.

We also lost six old rosemary bushes over the same two winters so I have replanted half a dozen more and taken cuttings for future reserves. The truth is that although rosemary is very hardy it is never entirely happy with our cold, wet winters and our clay soil. Like all Mediterranean herbs it thrives on really good drainage, poor soil and burning sun. The best way to ensure 'poor' soil is to add lots of grit.

Sage puts on a mass of new shoots in spring but it is essential to be ruthless to stop it getting too leggy by cutting back hard in spring – which means sacrificing flowers in pursuit of culinary delights. The best sages for the kitchen are the ordinary *Salvia officinalis* and the narrow-leaved *Salvia lavandulifolia*.

French tarragon (*Artemisia dracunculus*) is delicious with chicken but needs to be brought in each winter to ensure survival. Do not confuse it with Russian tarragon, *A. dracunculus dracunculoides*, which is hardy but no good in the kitchen.

Fennel (*Foeniculum vulgare*) seeds itself everywhere so feels like an annual but is in fact a perennial. In good soil it will become 2m (6ft) tall with fluffy fronds of aniseed-flavoured leaves that go well with all fish. We also collect the seeds by the handful in September and store them in a sealed jar until sprinkling on a joint of pork before roasting.

Marjoram (*Origanum vulgare*) is one of the most used of all summer herbs in our kitchen. We grow French oregano, *Origanum* x *onites*, which has green leaves, and golden marjoram, *Origanum vulgare* 'Aureum', and although they have slightly different strengths of flavour, cut and use them indiscriminately. Once established, they grow strongly and will need regular trimming to keep the foliage bushy and fresh.

I grow lots of parsley, creating temporary hedges to line borders but a good range of herbs can also be grown in a modest container.

Some annual herbs are always grown in the vegetable garden. Garlic can crop along with other alliums, basil with tomatoes, and dill, parsley and coriander in rows. A word on parsley, perhaps the most useful of all kitchen herbs. Give each seedling enough space to become a hefty plant with masses of foliage. This way it will last for months and give repeated pickings. Chives are easy to grow from seed although a clump can be divided in spring to make dozens of plants. Cut back to the ground every six weeks or so for the freshest shoots.

Beware of growing mint with other herbs. It will take over if given half a chance. Best to grow it in containers. We grow peppermint for mint tea, and spearmint and apple mint for cooking. Although it will survive almost anywhere, mint thrives in moist, rich soil so water regularly.

HERBS FOR SHADE

There is a dominant notion that all kitchen herbs like hot sunny conditions with well-drained soil. This is certainly true for rosemary, lavender, sage, thyme and the other well-known Mediterranean herbs, but there is a good, if smaller, selection of herbs that can be successfully grown in a border or pots that are largely in shade. However they will all do very much better if given some sunshine during the day. A spot that is shaded in the morning but gets sun in the afternoon or evening is ideal.

Sorrel is somewhere between a vegetable and a herb but is happy in the shade as long as it does not get too dry. It has a very metallic, sharp taste that is excellent with eggs. Mint is happy in shade, and as I have said, is the herb, above all others, that should be grown in a container of some kind, even if it is then – as we do in our herb borders – sunk into the ground so that it grows and blends in with the rest of the kitchen herbs.

Parsley will tolerate quite a lot of shade but again does best in rich soil. It is a common mistake to plant parsley too close together whereas 15–23cm (6–9in) between each plant will give the best results.

Lovage is a vigorous herbaceous herb that will grow in almost any conditions and has distinctive, celery-tasting leaves. Angelica is another big herb well-adapted to a shady spot as long as there is enough moisture in the soil to sustain it. Basil is surprisingly tolerant of shade as long as it has plenty of heat and rich soil. Like parsley it makes a much healthier, productive plant if planted with generous spacing – at least 15cm (6in).

Aphids

Aphids come in many different forms: green aphids on your roses; grey aphids on brassicas; black ones on your broad beans; and white ones at the moment on my clipped box. But aphids are as much part of the garden as you or a song thrush. They are an important part of the exhilaratingly complex natural jigsaw that is your garden and to know them is, if not to love them exactly, at least to tolerate them.

We have become accustomed to assuming that a garden plant is a tender, fragile thing only protected from the ravishment of a battalion of pests and diseases by ceaseless human husbandry. However most of my horticulture in my own garden is little more than encouraging and aiding every plant's natural health. A perfect plant is one best adapted for the place that it is growing. To try and force it into being 'better' is asking for trouble.

Aphids are usually blown into the garden, although as summer progresses more and more flying aphids hatch and will move themselves around at will. They are attracted to bare soil but will quickly move to find young, sappy growth. So roses grown in a carefully weeded border are prime targets, as are broad beans in a well-maintained vegetable garden. They will pierce the cell walls of leaves and stems to suck the sap. This will potentially weaken the plant. It will also leave a wound, which might enable disease and viruses to enter.

So far so bad, but in a healthy plant this is no more of a hazard than midge bites or cuts and grazes for you and I. A healthy plant heals itself. The aphids then excrete a sugary exudate called honeydew, which is ideal for fungal mould to grow on. The classic example of this is the black sooty mould on the upper surfaces of camellia leaves from the honeydew excreted by aphids on the undersides of the leaves above. Unsightly, but not a disaster.

I am certain that it is best to grow plants as 'hard' as is commensurate with good health. This means only ever directly feeding a plant that is visibly ailing. It means not adding too much compost or manure once you have established a good soil structure. Do not overwater, nor protect a plant from wind and weather any more than seems necessary. This means using garden compost sparingly to provide a balanced feed of complex and accessible nutrients that are sourced mainly from the garden – and therefore returning its own to it.

Given the right wind-free, warm, moist conditions with plenty of young growth to feed on, aphids can certainly increase their numbers astonishingly quickly. Each aphid will produce live nymphs that have their next generation already formed

inside. Young are produced at a rate of about 3–6 per day for several weeks. When aphid colonies become dense some wingless aphids will move off to find new places to produce their young. In autumn, males start to be produced and after subsequent mating the females lay eggs, which will typically overwinter in cracks in bark.

But, in a healthy garden, predators will immediately be attracted to the aphids and as the aphid population grows so will predator numbers until the two balance each other out. Ladybirds eat aphids and are attracted by the smell that comes from the honeydew. Hoverfly and lacewing larvae also eat hundreds of aphids before they pupate. Hoverflies are attracted to the garden by umbellifers of all kinds so a good assortment of dill, fennel, lovage, cow parsley, carrot, celery, parsnip and parsley all left to flower will provide sufficient inducement for aphid-eaters.

Remember that you need some aphids in order to have predators that will control the population. So killing off all your aphids will certainly result in an explosion of aphids later on. This is because if there are no aphids the creatures that predate them will have to move elsewhere or starve. Then a fresh wave of aphids will move into this virgin territory with all its fresh growth and there will be nothing to control them naturally. Chemical intervention simply enforces this boom and bust cycle.

A temporary infestation can be dealt with by rubbing them off with your hand, hosing them down with water or, if neither of these methods seems sufficient, spraying a weak soapy mixture that acts as a contact insecticide damaging the hard shells of the nymphs. Although better than using a chemical insecticide because it is not residual and wears off very fast, I would only use a soap spray under extreme circumstances, as it will also kill ladybirds and any other insects – just as I would only use the organically approved derris or pyrethrum for the same reasons. Although I have not used it myself I know that some gardeners use a solution of stinging nettles and garlic (not necessarily together) soaked in water as an aphicide. But any kind of spray (and at Longmeadow we have never used any kind of pesticide, organic or chemical) should always be a last resort. A garden without aphids is likely to be half-dead. Better to accept and enjoy life in all its astonishing variety.

Ladybirds are always a sign of health in a garden and should be encouraged – not least because they feed on aphids.

Lawn problems solved

Worm casts and moles are a sign of the excellent quality of your soil and must be tolerated, although I confess that moles make me roar with rage at the damage that they do. Worms become a nuisance only in autumn when their casts smear the surface but these soon go and can be brushed back into the soil. They do no long-term harm. Ants are becoming increasingly common, creating powdery fine little casts, but again do no harm. Just brush them back into the grass.

Many plants are classed as clover, but it is the white-flowering *Trifolium repens* that's the major lawn weed. It grows most vigorously on alkaline soils that are low in nutrients. The leaves grow from long, bare stems from ground level, which spread across the surface of the lawn. Clover's top growth can die back after a few hard frosts, but it can also remain green in dry summers. Successful clover eradication is only possible if you don't let the plant get established. Put your fingers beneath the plant and rip out the stems. Remove all scraps of stem, as any left on the soil surface will regrow.

Fairy rings and pale brown toadstools are caused by the fungus *Marasmius oreades* and occur as temperatures fall and moisture increases, which is why we tend to see them in late summer rather than spring. Fairy rings compete with the grass for moisture and nutrients as the feeding threads (mycelium) spread outwards from the point of infection. Typically there will be a stimulation of grass growth at the periphery of infection, causing the grass to turn dark green.

The usual cause is something organic rotting under your lawn such as a tree stump or root. Digging it up and removing it will reduce the supply of nutrients to the fairy rings, and removing the fruiting bodies will reduce the spread of spores.

Leatherjackets are the grubs of crane flies. They eat grass roots causing dead patches. The simplest way of dealing with them is to aerate your lawn well to prevent stagnant soil conditions. An alternative is to cover areas of your lawn with black polythene overnight. In the morning remove the polythene with any leatherjackets that have come to the surface at night encouraged to stay by the moisture beneath the polythene. It's a great breakfast for the birds.

Red thread disease causes patches of grass to become bleached and the red growths of the disease appear between bleached areas. Avoid scalping your lawn when mowing, especially if you garden on sandy soil. The healthiest height for grass is about 2.5cm (1in) – much longer than most people regularly mow down to.

Flat, curved chafer grubs eat grass roots causing patches to turn brown and die. Encourage plenty of birds into the garden that will be happy to eat them for you. Pull away infected grass and re-seed or turf.

The most harm that you can do to a lawn is to cut it too short. The grass will be a lot healthier if allowed to remain at least half an inch (1.5cm) long.

Nettles

Despite being a major weed in this garden with the seeds washed in by flooding every year, I would not be without them. And even though they grow where we do not want them and occasionally sting appallingly, they are easy to dig up and are a huge improvement to the compost heap.

Every garden should have a patch of nettles, as they are a major source of food for butterflies. They are also good food for humans, too. Spring nettles are a delicious vegetable and particularly rich in iron. Pick just the fresh tips of each stem and cook them like spinach. Cooking destroys any risk of stinging.

Nettles are an excellent source of nutrients, especially nitrogen, magnesium, sulphur, iron and phosphate. Soak 0.5kg (just over 1lb) of nettles in a 5-litre (1-gallon) bucket of water for two weeks. Use as a liquid feed – as a general-purpose plant boost in spring and early summer – diluting the resulting nettle liquid with 10 parts of water.

Do not be tempted to use a stronger mixture and use once a week at most. It can be applied to the roots of the plant or as a foliar spray. Remember that most plants do not need extra feed and over-feeding, especially young plants, will do more harm than good.

JUNE

I always feel that June answers questions that the rest of the year poses. Some of these are practical: 'how will this border look at its very best?' and 'how sunny will this corner be at the peak of the year?' or 'where does the sun rise on the longest day?' But the most satisfying answers are to the hardest questions, like: 'why do I garden?' or 'how does such a small patch of this earth give me so very much pleasure?'

I am sure that the secret of June is that it is not the peak of the garden's year or aspirations. The vegetable garden is still surprisingly empty at the start of the month and although a June border is always lovely, it never has the range of plants or colours that come along later in summer. The nights here can still be genuinely chilly – we have had frost on 6 June before now – and growth can be slow. 2011 was exceptionally cold and growth was minimal.

But, in truth, none of it matters. The British garden – and countryside – is at its very best. So much is still promised even though the garden is brimming over with beauty and sensual delight. There is as much as anyone could want and yet more in the offing. Although mature trees have all their foliage and the garden is in its very best finery, everything still has the freshness and inner glow of spring. Nothing is jaded. Nothing has yet been taken for granted. June is growing and becoming – with untrammelled energy – every moment a celebration.

Midsummer's Day, the summer solstice, is a real place, in the same way that New Year's Day or Easter is a meaningful place in the cycle of the year, and these holidays should be celebrated with as much energy and enthusiasm as they were by our pre-Christian, megalith-building ancestors. If the year is a range of hills then surely Midsummer's Day, on 24 June, is its summit and from then onwards there is an element of loss running through the days, knowing that they are shortening and that the slope is tipping gently down towards winter.

So, June must be savoured to the very last drop. My idea of horticultural heaven is to be weeding or planting as the light gently falls around me and to be able to continue with light enough to work until after 10pm – although weeds have been known to be planted and precious seedlings weeded in the half-blind rapture of the June twilight! I carry these few precious evenings with me for the rest of the year rather like a pebble in my pocket that I can touch, and they see me through the dark days of winter.

ANNUAL POPPIES

For all their massed effect of brilliant colour when growing in a field or on a bank, annual poppies have an individual delicacy that I adore.

They can be extraordinarily prolific. Each poppy flower will produce around 17,000 seeds of which around 3,000 will remain viable and dormant in untilled ground for at least a century before bursting into flower when the ground is disturbed and the seeds exposed to light. This tendency to wait until ground is tilled explains why poppies will grow seemingly from nowhere when a garden is cleared for the first time for years, sparked into germination by disturbance of the ground.

Breeders have harnessed its promiscuity to select garden strains with the red shifting through every shade, although the reddest petal with the blackest base is 'Ladybird' which shines as bright as any ruby as it catches the evening sunlight in the Jewel Garden.

Perhaps the most significant deliberate domestication of the field poppy came in 1880 when the Reverent Wilks, vicar of Shirley in Surrey, noticed a single field poppy in the vicarage glebelands with aberrantly white bases. He liked what he saw so duly collected the seed from it and eventually bred the strain that we know as Shirley poppies that all have white, rather than black, bases to the petals. They tend to be more delicate and subtle than the field poppy. Their great virtue is subtlety, although they will tend to revert to field red unless you carefully collect the seed of the paler, pastel shades.

All opium poppy seedheads look alike so I tie a thread around my favourites whilst they are still in flower so that they are identified for seed-collecting later in summer.

The opium poppy, *Papaver somniferum*, is my favourite poppy of any kind and as well as superb flowers has wonderful foliage (whilst they are young and fresh at any rate) and some of the most beautiful seedpods in the whole natural world. They grow in almost any conditions and will reappear year after year with complete reliability.

Its leaves are a fabulous glaucous grey, as lush as any lettuce, and the stems will grow to 1–1.5m (3–5ft) tall in rich soil but are so narrow it will squeeze in amongst neighbouring plants without swamping them. They look especially good with roses, but in truth you can hardly go wrong with them, whatever they accompany. The flowers can be any colour from white to a deep, deep purple and range from being single, fringed or arranged in multi-layered frills. My favourites are the peony-flowered types that have petals like crumpled ruffles of shredded silk. Others have cut fringes or create such a mass of petals that they are almost circular. 'Black Paeony' has an intense plum colour and, at the other end of the palette, 'White Cloud' is perfect for a white border. I also love the crested heads of the *Laciniatum* type that look like spectacular head-dresses for a day or so before they fully open.

Opium poppies are worth growing for their seedheads alone and unless you are collecting the seeds of a particularly attractive one, leave the plant in the ground simply for the urn-shaped fruiting heads. If you do want the seeds, tie a piece of twine around the stem as soon as it flowers (you think you will not forget which one it is but you always do) and cover it with a paper bag when the last petal drops.

Always sow annual poppies in situ as they hate being moved except when tiny. Either scatter the seed and let them grow where they land or sow them thinly in zigzags so you can differentiate the emerging seedlings from any competing weeds.

Although bleeding hearts take their name from the pinky red colour of the flowers, the white form, **Dicentra formosa** *f.* **alba**, *is a beautiful plant and ideal in the Walled Garden.*

BLEEDING HEARTS

Dicentra is part of the same family that produces corydalis and fumitory and is a close cousin to the poppy. There are 17 species of *Dicentra* with *D. spectabilis* by far the most common in British gardens, although it originates from China. All dicentras thrive in the cool shade of woodland margin – which in the garden effectively means under the canopy of any deciduous shrub or small tree, whereas if you plant them in full sun they will not flower so freely or long and their foliage will be less luxuriant, so give them shelter. We have them growing in the Walled Garden in the lee of a plum tree which seems to be about right for them. Like all woodland plants, they like an open, moisture-retentive soil so adding leaf mould as a mulch every spring will make them happy.

If you want blood colour from your bleeding hearts then 'Bacchanal' or 'Adrian Bloom' are the ones to go for, as they have crimson flowers. The white version, *D. formosa* f. *alba* is good and contrasts better with its paler, more glaucous leaves. There is a golden-leaved form 'Goldheart', 'Spring Morning' which has paler pink flowers, and 'Stuart Boothman' which has particularly delicate ferny leaves (with a slight mauve tinge to their grey-greenery) and flowers with a slight mauve tinge to their pinkery. However these are all refinements for the dicentra aficionado. The straight species is never less than good.

There is a climbing dicentra, *D. scandens*. It is herbaceous with flowers that can be white or yellow with touches of pink to the tips, and will continue in bloom for months in summer and autumn. It is not a showy, knock-em dead sort of climber but has a quiet charm, and although I have not grown it myself yet it is high on my shopping list of plants.

CLASSIC ROSES

The classic roses come to their peak in the Walled Garden around the middle of June and – unlike hybrid tea or floribunda roses – classic, or old, roses only produce one batch of flowers in summer. This is their hour.

I was first drawn to these plants by the romance of their names – both of the broad family groups such as gallicas, albas, damasks, bourbons and centifolias, and the individual roses such as cuisse de nymphe, 'Souvenir de la Malmaison' and 'Pompon de Bourgogne'. However as soon as I planted them and saw them grow it was the flowers, above all, that I adored. At this time of year every Saturday morning I go into the garden at dawn with a basket and pick a selection of roses. This might include the extraordinary overlapping petals of a centifolia like 'Chapeau de Napoleon'; the intense red of the gallica 'Tuscany'; the delicate blushing pink of the alba 'Königin von Dänemark' (Queen of Denmark); or the pure white of 'Madame Hardy' with her green eye at the heart of the ruffle of petals. I never tire of the sliced-off flowers of 'Charles de Mills' or the deep plum-coloured flowers of the hybrid perpetual 'Souvenir du Docteur Jamain'.

Their astonishing beauty would be reason enough to go to great lengths to grow these flowers, but the truth is that few plants are as trouble-free or easy to grow. Gallicas, albas and rugosa roses are all really tough shrubs and will grow in almost any soil or position. All roses do well in heavy clay soils and there are a number of superb classic roses like the deep red 'Souvenir du Docteur Jamain' and the pure white 'Madame Plantier' that we have planted against the east-facing wall of the Walled Garden because they will thrive in shade.

To get the best from any rose it is a good idea to dig the ground deeply, incorporating plenty of manure or garden compost. This is not so much to feed the plant but to improve the condition of the soil to achieve the balance of good drainage and water retention. Roses are thirsty and although they like sun, they never do well growing in dry soil.

All roses should be planted deeply in a generous hole so that the roots have room to spread. Sprinkle mycorrhizal fungi granules onto the surface of the hole (so that they come into direct contact with the roots) and a little more on top of the roots themselves. This seems to hugely help the adaptation of the soil's bacteria to the plant and enable it to take up all available nutrients. If you have large borders, planting in groups of three in one large hole is a good way to create one large shrub smothered in flowers. When you buy roses in a container the graft point or union is invariably 2.5cm (1in) or more above the soil level but this should be buried well below the soil level to reduce suckering.

Suckers are recognisable from the new shoots of the cultivar because they always appear from the roots, rather than above ground, and often a little way from the rest of the plant. They should be pulled out of the ground rather than cut as cutting only prunes them and stimulates more growth.

Finally, whatever type of rose you are planting, even if it is a large climber, prune it hard immediately after planting, down to about 15cm (6in) from the ground. This will encourage strong growth from the base of the plant and can obviously only be done at this point in the plant's life.

The silkily voluptuous flowers of the rose 'Chapeau de Napoleon' emerge from a wonderfully crested bud which, in its resemblence to a cocked hat,gives it its name.

Pruning and maintenance

If pruning is to be done at all I believe that the best time is in late summer, after flowering, and it is only necessary to remove any dead or diseased wood, cut back any exceptionally long shoots or any branches that are rubbing or obviously overcrowded. But, in general, most classic roses only need pruning in this way every two or three years.

Dead-heading is essential if you want to keep flowering going as long as possible. It is a form of pruning and rather than merely removing dead petals, cut back to the first leaf below the fading flower. However leave a final flush of flowers in July to let hips form as they both look good and provide food for birds in autumn.

Classic roses are strong, healthy plants and in general survive almost anything to flower profusely year after year. But all roses might show the following symptoms from time to time.

Balling is the state where the bud almost opens but the outer petals form a carapace that stops it developing into a flower at the last moment. The result is a ball of petals that rots and eventually drops off. It is caused by rain at the late bud stage followed by hot sun. This can be salvaged if you carefully tease apart the outer petals releasing the flower within.

Black spot is a fungal disease made worse by too much rain. The best policy is to prune in the New Year so that each plant has good ventilation, not to crowd the roses with too many leafy herbaceous plants and to collect up all fallen leaves and burn them so that the fungus does not linger in the soil at the base of the bush.

Powdery mildew is a pale grey mould on shoots and leaves, and is exacerbated by the base of the plant being too dry – which can happen even in a wet summer if the rain is light and never really soaks into the soil deeply.

EREMURUS

Eremurus, or the foxtail lily, is one of the more dramatic flowers that a garden can grow, rising on bare stems to make an elegant tower of flower. They are grassland plants, surviving by being unpalatable to grazing animals. They take a year or two to get going but by June they have added their dramatic, 2m (6ft) tall fuzzy spires to the flower border. They grow well in the Dry Garden where they get a good summer baking but I like them best in the early morning when the sun rises over the top of the yew hedge that closes the Walled Garden and the spires catch light and become gloriously incandescent.

All traces of the plants vanish over winter until the handsome basal leaves develop in spring. They have fleshy, rather brittle, spidery roots, rather like asparagus, that need very sharp drainage and as much sun as possible to do well. They should be put in just below the surface, with lots of grit beneath them and a shovelful more over the top. Given the right conditions they spread well and will self-seed, although they take four years from germination to flowering. They go dormant as soon as they stop flowering until the roots start to grow again in September, so the plants are best lifted and moved, if need be, in high summer.

HARDY GERANIUMS

Much of the charm of hardy geraniums, or cranesbills, comes from the ease with which they adapt to almost any situation in any garden. There are cranesbills from alpine scree, woodland, meadow and bog, so you will always find some that will thrive wherever you need them. I let them spread almost indiscriminately in my own garden – not least because they are easy to hack back – and I crop them flush to the ground. This tidies them up, if tidiness is needed, but also encourages a flush of fresh new growth.

The timing of this cut back, which applies to all early-flowering herbaceous perennials, is hotly debated – in our household at least. Early to mid-June is about right but if you cut back too hard too soon you lose a part of the lovely fullness and froth of early summer. But unless you cut hard in June you will end up with an even bigger hole at the beginning of August, which is when the regrowth really comes into its own. But whatever your precise timing, it is best to cut the more vigorous cranesbills back to create some space and to encourage a second flush of leaves and flowers in August and September.

Other than the need to cut it back once or twice a year, hardly any plant could ask less of the gardener than the cranesbill. They range in size from 10cm (4in) to 1.2m (4ft) and are available in all colours save yellow. I grow both *Geranium* 'Claridge Druce', which has candy-pink flowers, and *G. macrorrhizum*, which has flowers that vary through shades of darker pink and is more vigorous, for their ability to tolerate the bone-dry shade under the hazel tree in the Spring Garden. In autumn, the leaves of *macrorrhizum* turn a brilliant orangey-red, peppered with the yellow leaves falling from the tree above it. Lovely.

Two large herbaceous species are native to Britain, *G. pratense* and *G. sylvaticum*. *G. pratense* spreads wildly in a border, seeding itself with enthusiasm, so we happily hack this back as soon as the petals begin to fall, although it is never less than lovely with flowers that range from pale pink to blue. I have also just started growing it in the long grass of the Writing Garden as a perennial wildflower, reckoning that its vigour will stop it getting swamped by the competing grass. The double-flowering versions of *G. pratense*, 'Plenum Caeruleum' and 'Plenum Violaceum' are, like most double flowers, both sterile so can be left longer, although dead-heading with shears will prolong flowering. *G. pratense albiflorum* is a white version and 'Mrs Kendall Clark' has petals striped blue and pale grey which will appear very early.

G. sylvaticum is found under trees on hillsides in the northern half of England and Scotland and likes lush conditions, which is hopeful, despite our Midlands' position. There is a very blue variety, 'Amy Doncaster', which looks a goer. Another blue, *G.* 'Johnson's Blue' is ubiquitous but rightly so, as it is a tremendous performer, holding its violet flowers for up to ten weeks at a stretch in the Walled Garden.

Geranium endressii is good for smaller spaces and is one of the first to flower in spring. 'Wargrave Pink' is delicate and, I think, more charming than the slightly mauvish tinge to the species.

G. phaeum has spread right through the shade of the Spring Garden, its deep burgundy flowers contrasting with the cow parsley in late May. There are many cultivars of *G. phaeum*, ranging from almost black to white. In fact, cranesbills tend to

hybridise so if you wish to preserve the purity of a particular cultivar or species then do not plant them side by side or else you will end up with a cross that is likely to be a lesser, muddier version of either of its parents.

G. psilostemon produces flowers of intense magenta with black centres. It is another vigorous plant, growing to a height of 1.5m (5ft) and a spread of 2.2m (7ft). I love it and let it have its head until it takes over too much space. In fact, like most herbaceous plants, it needs dividing and replanting every few years to keep its energy levels up and the best time to do this is immediately after cutting back.

G. 'Ann Folkard' is a chance cross between *G. psilostemon* and *G. procurrens* and has flowers that are magenta influenced by purple. This is an important plant in the Jewel Garden and weaves its brilliant way through the borders linking purples, oranges and reds with astonishing success and fluency and it will sprawl either outwards or up through a shrub for 1.2–1.5m (4–5ft), almost like clematis. The leaves start out lime green, darkening with maturity so the growing tips fleck the entire plant throughout the season – which is miraculously long as it flowers from May through to September.

The hardy geranium 'Ann Folkard' weaves through the Jewel Garden and its combination of magenta flowers and lime-green foliage knits the borders together into an intense but harmonious tapestry.

The native honeysuckle, **Lonicera periclymenum**, *fills the lanes around Longmeadow with its fragrance and is the perfect link between garden and countryside.*

HONEYSUCKLES

One of my favourite joys of this midsummer season is the range of scents, and my favourite of all is the humble honeysuckle. Its fragrance saturates these sunken Herefordshire lanes, the warm pools of air soaking up the scent like blotting paper till you can almost see it. Yet it never becomes overpowering or crude in the way that jasmine or even lilies can. Honeysuckle is fruity and warm and gently erotic.

The botanical reason for this strength of smell is to attract the moths that pollinate it – hence its increased power at night. They can apparently detect it up to a quarter of a mile away. Our noses may not be that acute but the lovely scent of honeysuckle is enough to draw anyone into the dusk of their garden on a summer's evening and the lovely flowers earn their keep well enough in the brightness of the midday sun. A perfect plant – right around the clock.

The wild honeysuckle, *Lonicera periclymenum*, or woodbine, works best if it is in some shade, needing warmth rather than direct sunlight to bring out the best of its scent, and a west-facing urban wall is ideal. The main thing to avoid is letting it dry out too much. Its natural habitat is woodland and hedgerows with light dappled shade and rich but well-draining soil.

It is wonderfully good for all wildlife. The flowers start white but turn yellow after they have been pollinated by bees and eventually bear round red fruits that are important food for songbirds. The leaves are eaten by the larvae of white admiral and marsh fritillary butterflies and the tangle of stems makes excellent cover for nests.

Most honeysuckles flower on the previous year's wood so should be pruned – if necessary – immediately after flowering, however *L. japonica* flowers on the same year's growth, so should be pruned each spring. Mine is in almost permanent rather dry shade so the chances of it flowering are slim as the combination of too much shade with too little water will always inhibit flowering, but the plant survives this mistreatment with a degree of robustness. Nearly all honeysuckles will grow in shade but if the flowers can have sunshine for half the day they will be more floriferous and the fragrance will be noticeably stronger. To keep a garden honeysuckle in tiptop condition you must provide the roots with rich soil and plenty of shade and the flowers with some sunshine, preferably in the evening.

We have *Lonicera periclymenum* 'Belgica', or early Dutch honeysuckle, by the main door leading to the garden with its raspberry ripple flowers – white streaked deep pink – and good fragrance. The late-flowering Dutch honeysuckle, *Lonicera periclymenum* 'Serotina', flowers with a deeper purple splash to replace the early version's pink, and smells fine too, but later.

I am planning to plant *L. caprifolium*, sometimes sold as Italian honeysuckle. It is deciduous, copes well with shade and has fabulous scent. *L. japonica* 'Halliana' is another that I want to add to this garden; it is deciduous but hangs onto its leaves through all but the coldest winters and flowers right through midsummer. But it needs sun to perform at its best – and we are very limited in plantable wall space. However I rather fancy twining it up an apple tree in the orchard.

LAVENDER

Few plants evoke so many things so powerfully as lavender. It labels and defines a colour, even though, on inspection, lavender comes in lots of shades from dark purple to white. It also colours a whole mood or atmosphere of gentle refinement and prettiness and, above all, it evokes a fragrance that is unique and cannot be described in any comparative terms. It is curious how this most Mediterranean of plants has become so English, so perfectly suited to accompany tea on the lawn and a hazy patrician charm. The scent of it released by your fingers crumbling a few of the tiny flowers will trigger a chain of evocations that can be provoked by nothing else. The combination of toughness and softness and associations with summer afternoons and a benevolent sunshine that it carries with it, improves any garden. It is the ideal plant for a pot, architectural, resistant to almost total neglect and long-lasting.

Lavender in gardens tends either to be a loose hedge or a single plant whose softness is part of an almost archetypal prettiness. The secret of keeping a lavender bush in good shape is to clip it hard immediately after flowering, although do not cut into the old wood. With this treatment it will hold a tight ball well and is a much cheaper and quicker-growing alternative to box if you have a very sunny, well-drained site.

Lavender, like rosemary, hates sitting dormant in cold water, although that is not to say that it does not need regular watering in summer if grown in pots – and I have lost plants in pots through underwatering. Given the right conditions a lavender bush can live for ages, developing branches like a blacksmith's forearms, its flowers sparser and sparser in proportion to the woody growth.

There are many different types of lavender, with different varieties and different colours to choose from. *Lavandula angustifolia* is common or English lavender. It has the familiar mauve spikes of flower and will grow to a height and spread of about 90cm (3ft). There is a white form, *L. angustifolia* 'Alba', which does not grow quite so tall, and *L. a.* 'Nana Alba' which, like all plants with nana in their name, is small even when full-grown. *L. a.* 'Rosea' has pink flowers, as does *L. a.* 'Jean Davis'. But to my mind pink lavender is like white chocolate, perfectly nice but an aberration.

The two most common varieties that you will find in any garden centre are 'Munstead' and 'Hidcote'. 'Hidcote' is a deeper mauve and a bit more vigorous than the paler, bluer, faster-growing 'Munstead'. Both make good hedging plants, where hedge and edge combine to make a gentle demarcation line.

Lavandula stoechas, or French lavender, has mauve bracts on top of the flower spikes that start looking good rather earlier in the season than common lavender and narrow leaves that grow markedly up the stems. *Lavandula lanata* makes a dome of soft woolly leaves, which then throws up long spikes twice as high again, topped with purple flowers. I grow both of these in terracotta pots and bring them indoors in winter to protect the plants from the worst of the wet and cold.

L. dentata has leaves that are prettily ribbed or crimped and the flowers also are topped with bracts although of a paler, blue colour. It is not entirely hardy so needs protecting in a cold winter or bringing indoors. *L. latifolia* is an upright species, with broader leaves and you might find it sold as Dutch or grey hedge lavender. It is crossed with *L. angustifolia* to make *Lavandula* x *intermedia* which is sometimes known as old English lavender and this has produced perhaps the biggest type of lavender you can buy called *L.* × *i.* 'Pale Pretender' (which is also known as 'Grappenhall').

My granny would cut the flowers on their long stems and dry them in the airing cupboard before putting the dried flowers – which fell apart and looked like seeds –

in little muslin bags, stitched together with lavender cotton. She then put them in amongst her drawers, presumably to keep the clothes smelling fresh. Lavender flowers are best dried by cutting the stems just as the flowers open and placing them on open trays or hanging them upside down in bunches. The main thing is not to dry them too fast.

Lavender will grow from seed as well as take easily from cuttings. Sow the seed in autumn and overwinter the seedlings before pricking them out in spring and planting outside in early summer. They will then grow on fast. Cuttings are best taken as semi-hardwood cuttings in late summer from new growth, placing the cuttings in well-drained compost. Place the rooted cuttings in a cold greenhouse or coldframe over the winter and plant out the following spring.

PELARGONIUMS

Pelargoniums originate mainly from South Africa with 125 species from around the Cape alone – and I have seen them growing in the Fynbos above Cape Town as great shrubs flaunting pink flowers in that extraordinary subtle and beautiful landscape. When growing them – as with any plant – it is useful to remember their natural habitat and this bare, dry, rocky terrain.

The zonal ones are the familiar hothouse bedding plants with leaves like opened fans, often with a chocolate rim or centre. They hate getting wet, get mould, black leg and rust, but have a genuine, hot exotic aura that makes them worth growing. It is a good idea to grow them, either in pots or planted out, at the base of a sunny south-facing wall that will reflect heat and keep them dry. They are either propagated from seed, in which case they are treated as bedding annuals, or from cuttings, which produce larger flowers and will survive as perennials.

Unique types are shrubby plants with masses of small flowers as well as foliage that is scented when crushed. They mostly date from the turn of the nineteenth century – in other words the absolute heyday of the greenhouse and cheap labour – and flower for a long time.

Pelargoniums and lavenders flank the sun-baked steps in the courtyard. Terracotta pots always look good and are ideal for Mediterranean shrubby plants.

Regal types flower in early summer, the mass of flower with overlapping petals almost obscuring the foliage. They can have some of the best and richest colours of all pelargoniums, with 'Springfield Black' a deep burgundy and 'Dark Venus' a superb plum colour. They need more watering than other pelargoniums and a warmer minimum temperature in winter if they are to survive.

Ivy-leaved ones derive from *Pelargonium peltatum*, and whilst they often have masses of small flowers, their real virtue is that they trail and therefore are good for hanging baskets, window boxes and letting cascade down walls.

Species pelargoniums, like all species plants, tend to be tougher, less showy and often more interesting than hybrids that have been bred from them. My current favourite, in flower on a little table outside the back door, is *P.* 'Splendide' which has viola-like raspberries-and-cream flowers and a very unpelargonium-like glaucous leaf.

Quite a few of the species have scented leaves and are worth growing for this tactile fragrance, regardless of their flowers. We have them in pots flanking the warm, south-facing stone outdoor steps to my workroom. Every time I go up and down the medley of their fragrances sifts entrancingly through the air.

There is the cream variegated 'Lady Plymouth' which comes up smelling of roses, *P.* 'Graveolens' that has a delicious orangey fragrance, *P. odoratissimum* which smells of apple, *P. fragrans* which exudes astonishing pine freshness (although there are those that swear it smells of nutmeg), *P. tomentosum* that is pepperminty and the most famous of the lot, 'Mabel Grey' which is lemon-scented. By the way all these should be watered with rain rather than tapwater.

Finally there are Angel pelargoniums, which were bred from *Pelargonium crispum*. I have a small *P. crispum* and it is modestly handsome, the leaves growing tightly to the upright stems. But the Angels are not terribly typical of this, being looser and bushier and having the main advantage of flowering continuously all summer. I reckon that all Angel pelargoniums look pretty similar, if attractively so. Although there is one very special one called 'Sarah Don' bred by Roger Jones at Oakleigh Nurseries near Alresford in Hampshire and named directly in honour of my wife, Sarah. It has a golden variegated leaf and magenta and paler pink flower and is, like its namesake, magnificent.

We love pelargoniums of all kinds but the Angel pelargonium 'Sarah Don' has, for obvious reasons, pride of place.

Growing pelargoniums

As a general rule, the harder a pelargonium is treated, the better it will flower, because they only start to flower when the roots become constricted. So I keep them in smallish terracotta pots and do not feed them at all. However you can make any plant grow rapidly by repeatedly repotting it into a slightly larger container before it starts to flower. If you keep doing this, the plant will continue to grow vigorously until its roots become constricted. Then, when it is as big as you want it, leave it in the pot it is in and as the roots become increasingly constricted it will flower profusely.

If you have a pelargonium that has become too big and unwieldy and you are not sure how to prune it, it can always be cut cleanly across about a foot from the base, and will regrow vigorously.

Most problems with pelargoniums are due to overwatering. They are adapted to very harsh, dry conditions and need only minimal water, especially over winter. Let the plants completely dry out between each watering. If the leaves start to become tinged with orange or yellow, this is an indication to back off from the watering for a while.

Take cuttings from mid-August. Some, like the Angels, take very easily. None are difficult. Do not put the cuttings in a polythene bag (or mist propagator if you have one) but water in and then keep the tops dry without letting the compost dry out.

Cut them back to half their size in autumn and bring into greenhouse, conservatory or sunny porch. Keep dry and cool but frost-free. Most need a period below 10°C (50°F) if they are to flower the following season, so do not overwinter them in a warm room. A constant temperature between 5–10°C (41–50°F) is ideal. From March start to water them and as the temperature rises they will start to put on some growth. In early May cut them again to a good shape, repot to a slightly larger container and give them the sunniest spot you have and let them get on with it, watering no more than once a week.

THISTLES

All thistles thrive on the heavy soil at Longmeadow – although not all are welcome. The spear thistle, *Cirsium vulgare*, is a common weed in these parts, developing spines like needles. When young it can go on the compost heap but as soon as the spines harden up it must be burnt or finely shredded. Its close cousin the creeping thistle, *Cirsium arvense*, has a soft, sappy stem which has a habit of snapping off when you try to pull it up. It tends to get in under the hedges of this garden, spreading by lateral roots as well as by seed. We also get a lot of sow thistles, which are sappy and easy to pull up in well cultivated soil and then composted.

The other intrusive thistle into this particular garden is the burdock. It grows to at least (2m) 6ft and the burrs snag appallingly on any jersey – and the dog's fur – all winter long. The answer is to cut it at ground level as soon as you see it and to go on doing so until it weakens and dies. But it is a statuesque plant and part of folklore – the leaves were used to wrap butter in and the young shoots and roots used to make dandelion and burdock beer.

There are plenty of thistles that deserve an honoured place in a border. *Cirsium rivulare* 'Atropurpureum' is a plant that delights in boggy conditions or heavy clay soil and is very happy in our Grass Borders. It grows to about 2.2m (7ft) tall with flowers that are richly plum-coloured and the leaves hardly prickly at all, which makes it a very border-friendly plant. It does have a habit of suddenly collapsing and not reappearing the following year and is sterile so will not produce seedlings, so it is a good idea to lift it every couple of years and take some root cuttings.

The globe thistle, *Echinops ritro*, is a tough herbaceous perennial, happiest in poor soil as long as it gets some sun, and although its leaves are horribly prickly, the pom-pom blue flowerheads justify the occasional painful brush with them. But it rapidly makes a huge clump and needs rigorous reduction every year or two. But I love its perfectly round mauve heads that have a weirdly functional moment just before the buds open, when they look spectacularly unplant-like, resembling steel tooled to industrial specification. *Echinops bannaticus* 'Blue Globe' has darker blue flowers and those of *E. bannaticus* 'Taplow Blue' are a more intense blue. *E. exaltus* is huge, growing to 2.2m (7ft) with silvery white flowers.

Perhaps the most popular thistle is the giant sea holly, *Eryngium giganteum*, although it is not really a thistle at all but a distant cousin of the carrot – although its armoury of spikes and prickles is certainly very thistly. It is commonly known as Miss Willmott's Ghost because Miss Willmott apparently went round secretly –

and rather irritatingly – scattering its seed in other people's gardens. Nevertheless it is a good plant, distinctly silver, tinged with the blue that so marks sea hollies and leaving a dried husk of itself all winter with a seedhead not unlike a teasle, *Dipsacus fullonum*, which is also another incredibly spiky plant and one which loves wet ground. *Eryngium giganteum* 'Silver Ghost' is smaller and more silvery white. It is a short-lived perennial, which means that it is lucky to flower twice, and like all eryngiums it prefers well-drained, poor soil.

ARTICHOKES

Artichokes are one of the few vegetables that are truly worth their place in a border as well as producing delicious food. They are easy to grow and a couple of plants will provide a number of delicious summer meals. They do best in full sun and rich soil although they need good drainage. A generous addition of compost in the planting hole deals with both requirements.

It is best to be patient and remove any flower stems in the first year, allowing all the growing energy to go into the roots and leaves. Then the second year it should grow with much more vigour and produce at least two cuttings of chokes. Although they are perennials and can live for many years, after their third year of maturity (fourth year from seed) the plants become less productive and are best ditched.

But if you take offsets from the second year onwards you will have a constant supply of second and third year plants – which will produce the best harvest. For an offset, simply chop off a piece of root in spring, with a sliver of leaf attached from the parent, and plant it where it is to grow. Do not worry when the leaf shrivels and dies – which it surely will – as new ones will take its place.

I have 'Violetto di Chioggia' growing in the Ornamental Vegetable Garden which has marvellous plum-coloured chokes that look as good as they taste, as well as 'Green Globe' which produces a large round choke with flat, smooth overlapping plates.

Like alliums, thistle heads are made of a compact globe of individual florets that, on close inspection, form a lovely abstract geometry.

The Top Veg Garden in midsummer, doing well – especially considering that five months before this photograph the site was dock-infested grass.

PARSLEY

Whereas some herbs are attractive and useful sideshows, some are worth the space, time and attention of any vegetable. I would include garlic in this second group and most definitely parsley.

Parsley is not difficult to grow so needs no extra skills and grows reasonably fast so demands no special allocation of time. What it does deserve is plenty of space.

There is nothing wrong with parsley in a pot per se. The affront comes from the relationship of plant to pot and the cavalier disrespect that it so often shows to this prince of herbs. Go into any supermarket or garden centre and the chances are that you will have the opportunity to buy parsley in a little pot. But this will not give you anything like the potential that even a moderate understanding of the plant would give you.

This is because if you look closely you will see that your parsley is comprised of perhaps dozens of weedy, etiolated seedlings giving the appearance of munificence. But each of these little plants is desperate for space. They are like battery hens, producing the goods but for a tiny period and under appalling conditions. Take a snip or two of your herb and the whole thing gives up the ghost. This means – and perhaps one might be justified in being a teeny bit cynical here – that you have to go back and buy another pot from those nice folk at the supermarket.

Grow some yourself from seed and each tiny seedling will last from 6–9 months and give a harvest more bountiful than the biggest pot that mass production will supply. Grow lots of seedlings and happiness, health and a great deal of financial saving is guaranteed.

Parsley is not a Mediterranean herb and thrives in well-drained but fertile soil, with plenty of moisture. It will also tolerate some shade. I have found that it sits much happier as part of the vegetable rotation than trying to fit in with the needs of the herb garden (which is based around Mediterranean herbs with their desire for poor soil and maximum sunshine).

It is a biennial and a member of the carrot family and although it can be fitted in anywhere – I often use it as an edging plant, not least because it makes it easier to pick – it is fundamentally part of the carrot, parsnip and celery rotational group. Each plant will develop a long taproot and a healthy parsley plant between three and nine months old should be 15–30cm (6–12in) tall and bushily frondescent.

Because it has the space to develop its true root system it will regrow a new flush of leaves whenever it is cut, again and again for the first nine months of its life, until the urge to develop its flowers becomes overwhelming and then it has to be dug up and added to the compost heap. When you do this you will be astonished at the size of the root.

I make two or three sowings a year, the first in January, then May and finally August. This ensures a limitless supply right through the seasons.

I principally grow flat-leaved parsley, which I think has a better flavour and texture, although the curly-leaved is good and decorative, too. But the flat-leaved kinds used liberally with vegetables, stews, soups, as part of salads and, with walnuts, as a delicious pesto for pasta or potatoes, is magnificent.

PEAS

It has to be admitted that bought frozen peas can be very good indeed. But if you grow them yourself you gain a whole range of experiences and sensations above and beyond reasonable tasting peas.

First there is the aesthetic value of peas growing up brushwood pea sticks. These can be any of last winter's shrubby pruning material although I like the traditional hazel pea sticks, which are the off-cuts and tops from bean sticks. Then there is the fabulous pleasure of picking a pod, cracking it open with thumb and finger and scooping a row of peas from the shell straight into your mouth. These raw peas have a sweetness and depth of pea-taste that money cannot buy nor technology replicate. Finally there is the slow, thoroughly hassle-free, unmodern pleasure of sitting with a glass of whatever you fancy with a bowl of freshly picked peas and podding them one by one, preferably along with someone else.

Whilst the magic of clicking open a ripe pod with your thumbs and pushing out a line of small, tender peas is one that should not be passed over by any gardener, sugar-snap and mange tout peas have their place too, and can be very delicious. It is a textural thing really and there is room for both in one's kitchen if not in the garden. My own feeling is that if you are short of room then go first for the real, full-podded thing.

Growing peas is easy enough. They like neither hot, dry weather nor cold wet soil, so the best time is in April and May, although it is worth trying a late June sowing. I grow 'Feltham First' (early), 'Douce Provence' (early), 'Hurst Green Shaft' (maincrop), 'Carouby de Maussane' (mange tout) and a wonderfully tall, fully 2.5m (8ft), and decorative purple-podded variety 'Ezetha's Krombek Blauwschok'.

Give them ground that has been manured that winter. I sow mine in double rows, peas spaced about 10–12cm (4–5in) apart in each direction and plenty of room between the rows to walk down and pick the peas. They should be given support as soon as they germinate so that the plants can twine up from the outset. The pods will appear as the last flowers fade and should be picked as soon as the peas are large enough to shuck.

When the bines start to turn yellow and the pods become wrinkled and hard, pull the entire plant up (making a good addition to the compost heap), fork the ground over lightly and use the site for winter brassicas such as cabbage, broccoli or kale which will gain much from the nitrogen that all legumes add to the soil.

(Previous page) A purple pea pod emerging from its flower and looking as exotically beautiful as any orchid. Although the varieties in the Ornamental Vegetable Garden are chosen to look good, they are all eaten – as was this pea!

(Opposite) Some pea varieties, like these purple-podded 'Ezetha's Krombek Blauwschok', can grow hugely tall and need really strong supports up to 2.5m (8ft) high.

Netting the strawberry bed in the Soft Fruit Garden. Blackbirds love strawberries and will strip the entire crop if it is not covered as soon as the fruits start to ripen.

STRAWBERRIES

I spent the years of 7 to 16 at boarding school and on my birthday, at the end of the first week of July, my parents would always visit with the special treat of strawberries from the garden at home. The strawberries came as reliably as the birthday but the quantity varied enormously from year to year. Some years I was presented with a small punnet with each rich red berry a rare treat and other years there were baskets of them that could be shared out to practically the whole school.

I accepted this for what, even at that tender age, I knew it was – the vagaries of the season and the weather. This lack of dependability did not in any way reduce the pleasure. In fact it made the strawberries we did get all the more delicious. However, although the strawberry season used to be a brief intense affair, nowadays we are saturated with them. Every day of the year you can eat fruits that look like a strawberry, smell just like them but taste – well mostly of a mushy nothing. And just as a birthday every day would soon pall, so do punnets piled high of cheap, bland strawberries.

But none of this is the fault of the strawberry. They can still be completely delicious if you grow your own. Only then can you appreciate a really proper strawberry, eaten warm from the sun (chilling removes all but the crudest flavour) and accompanied by single cream or with – and this is a bit unlikely but worth trying – a sprinkling of white pepper (the heat of the pepper brings out all the flavour of the strawberry).

The maincrop strawberry starts to fruit in June and is all done by the end of July or earlier. It is essential to choose your varieties with care and avoid at all costs the commercial types like 'Elsanta' that are chosen for their robustness in handling

and storage and aroma – which promises a lot but invariably fails to deliver. We have the French Provençal variety 'Gariguette', which is long to the point of being almost cylindrical, in the Ornamental Vegetable Garden, and is classed as 'mid-season'. The point about a strawberry like 'Gariguette' is that you are unlikely to buy it because it does not keep or travel well. The fruits are only at their best for about a day or two. If ever there was a reason to grow your own then this is a perfect example. I also grow a pink-flowering F1 variety in the Ornamental Vegetable Garden called 'Tarpan' which seems to offer more from its flowers than its fruit but certainly the flowers are distinctive and pretty.

Perpetual or remontant strawberries ripen between July and October, taking over when the conventional summer crops have finished. 'Mara des Bois' has proven to be exceptionally good for us from this group and it is rightly acquiring a high reputation for cropping well and having exceptional flavour.

The small but delicious Alpine strawberry, *Fragaria alpina*, fruits constantly from June to October. This is easily grown from seed and makes a good decorative edging, although it can become a bit invasive. However, if you are to have weeds at all then none taste better than this. I have some *F. vesca* 'Mignonette' plants which grow strongly for an Alpine and produce some of the most fragrant fruit of all, and 'Four Seasons' has proved to be healthy and very free-fruiting although the fruits could do with more sun than the poor things got in the miserable summer of 2011 for full sweetness and flavour.

In fact, most strawberries grow well from seed, and Alpines can only be propagated this way, but it is easier to propagate by taking runners from maincrop varieties. You can simply peg down the tendril either side of the immature plant and about a month later cut the stem either side of the pegs, and lift the rooted plant. To avoid transferring potential disease or viruses, it is best to take runners from one-year-old plants, after their first harvest.

Strawberries grow best on rich soil but whatever soil you have the addition of garden compost or manure will improve both the quality and quantity of the harvest. The best time to put in new plants is late summer, which will give them time to establish so that they produce a crop the following June. The young plants

We allow Alpine strawberries to spread more or less wherever they want because they have tiny, delicious fruits that are produced for months.

should be put in at 45–60cm (18–24in) spacing in rows or blocks with the base of the central crown at soil level. This will look ridiculously far apart but they need the room to grow healthily and to bear maximum fruit.

Strawberry plants are at their peak in their second and third years so it is best to dig up plants after their third harvest and put them onto the compost heap. Hence the need to take runners. As well as giving a diminishing crop after their third year strawberries are very prone to viral diseases that can remain in the soil for a few years so you should never replace old plants with new on the same site but keep a three- to four-year rotation on separate plots.

The fruits should be kept off the ground to avoid slugs and rotting, so tuck straw or a mat around each plant. Blackbirds love strawberries and are drawn to them like magnets as soon as they begin to ripen, so a net is essential once any of the green fruits start to turn red. This can be a temporary affair, held in place by canes and upturned plant pots that can be easily rolled back for picking.

To avoid a glut of fruit all arriving within the same few weeks you can encourage some plants to crop earlier by protecting them with cloches as soon as they come into flower in May. This will speed up the formation of fruit and the ripening process by a week or two. But keep the ends open unless it is very cold to allow insects in to pollinate and to keep good ventilation around the plants.

Once the last strawberries have been collected cut the foliage off to within about 8cm (3in) above the crown. New foliage will appear before autumn but most energy will go into the roots ready for next spring's regrowth. A mulch of garden compost after this annual cut back is the only feeding that strawberries need.

TAKING STRAWBERRY RUNNERS

When the strawberry plants stop producing fruit they send out vigorous new growths with foliage spaced out along their length. These are 'runners'.

Secure the runner either side of the first set of leaves with bent wire so that the stem touches the soil of a small pot of compost. Cut the runner beyond the pinned section.

When new growth is apparent – after about 2–3 weeks – cut the runner free from the parent so that it is now a small independent plant.

Transplant the runners to a new site that has not grown strawberries for three years. They should have a healthy root system and quickly establish as strong plants.

Cultivating bees

We all love honey but in the world of packaged food products it is not significant financially, worth just around £20 million a year in the UK. But honey is just a delicious by-product of the honey bee. Far, far more important is the role the bee plays in pollination. It has been estimated that 80 per cent of the western diet is dependent upon pollination by bees. Without bees the human race would rapidly starve and probably become extinct. In short, our lives depend upon bees as much as they do upon sunlight. Their steady decline is therefore a cause for real alarm.

The cause seems to be a combination of things. The varroa mite has become a major pest, attacking bees at every stage of their life cycle. It sucks the blood of adults and weakens them. In turn this makes them more susceptible to viral infections. Add to that the increasing evidence that agricultural pesticides are harming bees and you have a potentially disastrous situation.

But gardeners can play an important role in rectifying this situation. We can actively nurture and conserve the British bee population. Gardens are always a rich source of food for bees and with a little care can be made even more hospitable for them without any trouble or loss of pleasure to the gardener.

It goes without saying that you should not use pesticides. Create a garden rich in plants, allow a little gentle disorder and enjoy the privilege of hosting a vibrant bee population.

Bees do best in an environment where there is a range of plants providing a steady supply of nectar throughout spring and summer, with an intense burst of flowering at critical periods – like an orchard in May or heather on a moor in late summer. Bees will focus their pollen collecting on an area as local as possible to the

hive and will return again and again to a source of nectar for as long as it is available. So plants with a long flowering period and a succession of blooms are better for them than a short, spectacular harvest.

Honey bees like saucer-shaped flowers that are easier for their relatively short tongues to dip into. In a trial held in 1997 at Cambridge Botanic Garden, it was found that they had a distinct preference for some plants over others. The five most popular to bees were, in descending order of popularity: musk mallow (*Malva moschata*); common mallow (*Malva sylvestris*); lesser scabious (*Scabiosa columbaria*); cornflower (*Centaurea cyanus*); and wild clary (*Salvia verbenaca*). Hollyhocks and evening primrose were also high on the list. It is a lovely collection of plants quite apart from the enticement for bees.

Bees also love all fruit trees – in fact any flowering trees – and all legumes such as peas, beans, clover and sweet peas, as well as dandelions, blackberries, asters, ivy, and willow. Remember that bees need pollen and the smaller flowers of unhybridised species are likely to be a much richer source of this than huge show blooms on plants that are the result of elaborate breeding.

Make sure that there are shrubs, hedges or small trees to provide windbreaks and shelter. Leave sunny sites more open and ensure that they are filled with nectar-rich flowers for as much of the year as possible. If you notice a flight path of bees – and they make fascinating study – then do not block it with plants or objects.

Wet!

We live on a flood plain and Longmeadow often floods and the section that we call the Damp Garden can spend weeks under water, especially in winter although we quite often get flash floods in summer, especially if it has been very dry and the ground is hard.

The biggest lesson that I have learnt in gardening is not to fight the soil or the climate but to plant intelligently to make the most of both. It is elementary common sense but it always amazes me the extent to which people will go to buck nature. The clever thing to do is to go with the flow.

Soil can be improved to cope with excess rain with lots of organic matter dug into it as well as horticultural grit or sharp sand. This will open the soil out and make it easier for roots to work through and for water to drain. Likewise, very sandy soil needs organic material added to it so that water is retained a little longer giving roots more chance to access it. This also lets more oxygen in because compost will stop the light soil 'slumping' and closing up after heavy rain.

Any plant that has to cope with waterlogging is adapted to living with a very low level of oxygen, however the roots of many plants will deteriorate fast if waterlogged. The first signs of a waterlogged plant will be a yellow or orange tint to the leaves and the plant will wilt as though suffering from drought. It is perfectly possible to drown plants and some are far more sensitive to this than others. In my own experience the Mediterranean herbs such as rosemary and thyme are particularly ill-equipped for sitting in wet winter ground – yet until a few years ago many people were responding to the fears of global warming by recommending an increase in Mediterranean plants. However it does seem that climate change is likely to affect the UK as much by condensing rainfall into shorter, more intense periods, especially in spring and autumn, meaning that our gardens veer between increasing extremes of wet and dry.

There are fortunately a lot of perennial plants that do well in mild, moist conditions, feeding the need for fast, lush growth. Of course there is a fine line between true bog plants and plants that merely do well when the ground is wetter than normal. A bog is really permanent mud and whilst it can sometimes seem that way here, heavy or damp soil can dry out to a concrete-like hardness.

JULY

I love July most for its harvests. If the word harvest is redolent of cornfields and swathes of countryside, an allotment or vegetable patch is the nearest that most of us get to it. But however modest, it is a real and meaningful harvest, and a magical word bringing with it all the resonance of fulfilment and completion. Around the time of my birthday – the beginning of July – the first harvests of the vegetable garden begin.

First of all are the new potatoes, emerging white and surprisingly gleaming from the soil, almost blinking in the sunlight. No potatoes ever taste as good as this first shy crop of the year. Next is the garlic crop that has been growing since October. I harvest ours when the leaves start to fade and incline to set a minaret of seed. I lift them carefully with a fork so as not to damage the necks before drying them in the sun for a few weeks so that they will store through to the following spring. Beans, peas, the first tomatoes, all the salad crops, of course, tumble from garden to kitchen with an increasing flow. It makes supermarket shopping seem a drab, dreary process. All this bounty unquestionably tastes best eaten outside in the garden in the warmth of a July evening as the light gently fades.

It is not just the edible harvest that July parades, as the borders brim with the bright energy that characterises the flowers of the first half of summer. Best of all are the roses, which are at their peak at the beginning of the month, and for many of the most beautiful old roses it is a final display before they gradually recline into hips and thorny stems. It's a brief harvest, but incomparably beautiful, and worth waiting all year to celebrate.

July is also the month when we plant out most of our tender annuals and bulbs that have been raised from seed, designed to provide colour and drama in late summer and autumn. If the beginning of the month still belongs to the bright peaks of midsummer, by the end the garden has a new shape and palette, bigger, fuller and richer.

Of course, the nicest way to combine the fruits of flower and field is to eat outside as often as possible, and even in a summer short on real warmth or long sunny days, there is always some time in July to sit eating and digesting a meal, enjoying a glass of what most tickles your fancy, and watching the dusk wrap itself gently around the garden.

CLIMBING ROSES

I am very aware that we do not have enough climbing roses in this garden. When I visited the amazing garden at Ninfa just south of Rome in early May there were hundreds of climbing roses all in full bloom smothering every wall and climbing every tree. It was one of the great garden experiences of my life.

I realised two things. The first was that you can almost never have too many climbing roses – or any other flowering climber for that matter. Vertical colour is nearly always a good thing. The second was how magnificent many roses look if allowed to grow as big as they really want to rather than being constantly curtailed and clipped to stay within the limits imposed by our back gardens. So if you have room on your house or a suitable tree (an old apple tree is perfect) try growing a rose up it so that it can grow untrammelled. But even if this is not practical, then pergolas, arches, fences, walls and bowers are all ideal for training a rose over.

There are two types of climbing roses, climbers and ramblers. 'New Dawn', 'Albertine' and 'Dorothy Perkins' are climbers while 'Rambling Rector', 'Wickwar' and 'Paul's Himalayan Musk' – which we have in the Walled Garden – are ramblers. The major difference between the two is how and when they flower. There is a simple rule. Ramblers only flower once for a maximum of six weeks in summer and they tend to have many more and smaller flowers than any climber. Many ramblers are also very vigorous and one or two, such as 'Kiftsgate', can reach a truly enormous size. A single plant can easily smother an entire house, although many years ago I had a client who wanted me to design her small town garden and insisted on buying and planting no less than six 'Kiftsgate' roses along a short stretch of trellis along the edge of her roof garden!

They also produce their flowers differently. Climbers form their flowers on new wood that grows in spring before flowering whereas ramblers do so on the growth that took place in summer and autumn after flowering. This is why you often see a mass of hugely vigorous fresh growth on an unpruned rambler in late summer whereas by August many climbers are beginning to look a bit weary – as well they might after a long summer's flowering.

It follows from this that the two types of climbing roses need very different pruning regimes. Put very simply, ramblers should be pruned after they have flowered and climbers before they flower.

The best time to prune climbers is in late autumn although they can be done as late as March or April. The idea is to create and maintain a two-dimensional framework of about five long stems trained laterally and tied firmly to wires (or some other support) with side branches breaking all the way along these main stems. It is these side branches that grow vigorously in spring and carry the flowers. Ideally a third of the plant is removed right to the ground each year – the oldest, woodiest stems – so that it is constantly renewing itself. Left unpruned a climber typically becomes very bare at the base with a cluster of flowers right at the top or at the edge of an impenetrable tangle. If you do have such a tangle then the best thing is to prune it all back very hard and start again with the subsequent regrowth and in two or three years you will have established a good framework.

Ramblers are even more prone to create an unholy – and very thorny – tangle and this is often a result of trying to restrict too vigorous a plant in too small a space. If you wish to train a rambler to a specific shape then it needs careful control from the outset, and it should be pruned back to a bare framework in summer so the subsequent new wood carries a good display of flowers the following summer (without becoming overwhelming). Otherwise the best option is to plant it up the side of a tall tree and simply let it grow.

(Following page) The rambler rose 'Paul's Himalayan Musk' sprawls and twines through the damson trees at the back of the Walled Garden.

My favourite climbing roses

(all climbers unless stated)

For north- or east-facing positions (ie some shade):

- 'Madame Alfred Carrière' (pink)
- 'Souvenir du Docteur Jamain' (rich burgundy)
- 'Zéphirine Drouhin' (pink)
- 'Madame Plantier' (white)
- 'Madame Legras de St Germain' (white)
- 'Climbing Iceberg' (white)
- 'Aimée Vibert' (white)
- 'New Dawn' (pale pink)
- 'Louise Odier' (pink)
- 'Mme Isaac Pereire' (pink)
- 'Kathleen Harrop' (pink)

For south- or west-facing walls (ie sunny positions):

- *Rosa banksiae* 'Lutea' (yellow, rambler)
- 'Guinée' (rich red)
- 'Gloire de Dijon' (pink)
- 'Constance Spry' (pink)
- 'Bobbie James' (white, rambler)
- 'Rambling Rector' (white, rambler)
- 'Wedding Day' (white, rambler)
- 'Sander's White Rambler' (white, rambler)
- 'Cécile Brünner' (pink, rambler)
- 'Paul's Himalayan Musk' (pink, rambler)

Apart from black spot and powdery mildew (which are described in detail in June) there is another disease to which roses are prone. Rust thrives in wet, warm weather. It starts with small orange pustules on the underneath of leaves that can spread very quickly over the whole plant. These spores turn brown and eventually black, killing the leaves. Remove any leaf that has a trace of rust and burn it. As with all fungal diseases, clean up round the base of the plant scrupulously in autumn so that the spores cannot overwinter in the ground and then mulch generously in spring.

DAMP GARDEN

There is a triangular section of Longmeadow, surrounded on all sides by high hedges, which we call the Damp Garden – but in reality it is only damper than elsewhere for a few weeks a year when it is covered by flood water.

This deluge brings two things in its wake. The first is a deep layer of lovely silty loam that has built up over millennia. This never dries out yet never becomes boggy, and guarantees magnificently lush growth. Even when planted with nothing but common or garden natives it quickly becomes an exotic jungle. The second is a regular rash of weed seeds, dominated by nettles but also lesser celandine, Indian balsam, comfrey and lesser hemlock, all of which grow in that silty loam with astonishing enthusiasm.

It evolved into its current form gradually, as much as a result of the hedges maturing as any careful plan. It was, in fact, the very first piece of garden that we planted. I remember coming here in spring 1992, before we moved in, and planting some willow cuttings and primroses on what seemed to be a slight grassy bank running down to the boundary fence. The children were tiny and when we had finished this token horticulture we had a picnic with a rug, tea from a thermos and fairy cakes, the dogs earnestly eyeing the crumbs. Within just a few years the willows grew monstrous and needed a digger and stump grinder to remove them, and the primroses were dug up and shifted, first to the Spring Garden, and then to the Coppice where they have all regularly been divided and moved – but perhaps the same plants are still flowering there? Those dogs are long buried in the Coppice, too and of course, the children are all grown up.

It then became a wild garden and for a year or two was poised with a controlled anarchy that was enormously pleasing, but after a few years this became a weed-infested guilty space.

So, I put the compost heaps here and this worked well for a few years. But when my hostas outgrew their home in the Spring Garden I decided to re-house them in this under-utilised and under-appreciated damp bit. I moved the compost, dug out all the weeds, made a path, dug up and moved the hostas and arrived at what we have now: a garden devoted to those plants that do best in moist soil, some shade and which are robust enough to compete boldly with the inevitable influx of weeds.

(Previous page) The Damp Garden has a lushness in summer that is sustained by its long winter soaking which means that the soil never dries out.

GERANIUM 'ANN FOLKARD'

At this time of year I am busy cutting back the geraniums growing in the borders both to stop them smothering their less vigorous neighbours and also to encourage a second flush of flowers next month. But I leave the lovely 'Ann Folkard' well alone and let it ramble untrammelled. It is an indispensable addition to our garden, both for its foliage and its intense almost-magenta flowers (it is as bright a purple as that colour can possibly be, see page 149).

It was a chance cross about 30 years ago between *G. psilostemon* and *G. procurrens.* From the former it has inherited its intense flowers and from the latter its propensity to sprawl all over the place. It will work its way either outwards or up through a shrub for 1.2–1.5m (4–5ft).

The strikingly yellow-green leaves darken with maturity so the growing tips fleck the entire plant throughout the season. It stays in flower from the end of May until the end of September, by which time there are barrows of foliage to cut back and take to the compost heap. The plant retreats to a smallish crown which means that you can plant bulbs like crocus, daffodil and tulip close around it without them being swamped, as it starts its expansion after they have flowered. You can propagate it from cuttings, although these can be tricky to overwinter. It is easier and more reliable to dig up a crown in spring and cut it into divisions with a sharp knife.

LATE CLEMATIS

The late-flowering or group three clematis start to flower in the Jewel Garden in July and will, in some cases, continue right into October. No other flower can provide so many floral jewels for as long as a clematis.

We grow them up wigwams of long bean sticks cut from hazel but they can be trained up any surface from a wall or fence to a neighbouring shrub. What makes the late-flowering clematis so useful for a mixed border is their habit of only flowering on new growth. Their natural habitat is woodland and they grow in the shade of bushes and trees, reaching towards the light as they get taller and taller. But if they are cut back to the ground each year they will make extra vigorous growth and be covered with flowers from the base to the top. Thus, however big and sprawling they grow in the summer, all the top growth is cut back hard in late winter and the process can start again, without getting in the way of neighbouring plants. This annual clear out also makes them perfect for free-standing trellis as the weight of woody growth never accumulates enough to topple it over.

All the late large-flowering clematis are the result of crossing *C. languinosa* and *C. viticella* with subsequent breeding between the resulting offspring. All are deciduous and share the characteristic of flowering on current season's growth. Hence the need to remove all of last year's wood each spring. The ritual of this annual prune has become, for me, a celebration of the end of winter, regardless of the cold weather that is bound to follow.

None of these hybrids produces particularly attractive seedheads – unlike the late-flowering species. All the clematis in this group are tough and will stand any kind of winter weather that this country can conjure up.

Probably the best-known is *Clematis* 'Jackmanii' which was the first clematis to provide large purple flowers. You will often find *C.* 'Jackmanii Superba' at garden centres, which has even larger flowers with streaks or ridges of claret like scarification lines running down each 'sepal' (the bits that look like petals – and if that wasn't confusing enough, some books call them 'tepals').

Clematis viticella is the star of this group. It was introduced from central Europe in the sixteenth century and has since been bred to produce much-loved cultivars such as 'Madame Julia Correvon', 'Etoile Violette' and 'Polish Spirit' but the best is the plum-coloured, double-flowering *Clematis* 'Purpurea Plena Elegans'. This is a mouthful for a flower that is delicate in a fumbly sort of way, flowering against the

backdrop of its foliage like bejewelled buttons. It is a long-established favourite that has been grown in British gardens since Elizabethan times.

Although the late-flowering clematis have a huge assortment of flower forms, they all have an intensity of colour, as though it has been soaked into the deep velvet of the sepals, in shades of violet, purple and maroon. Some like 'Honora' change colour as they go along. It starts out a twisted funnel of plum and matures into a cheerful gappy purple flower. 'Gipsy Queen' holds its purple depths until it drops and 'Perle d'Azur' makes a good fist of being blue. I cannot get enough of them, and am always looking for another site to plant another variety so it can scramble amongst the rest of the late border flowers, adding its own inimitable celebration of summer.

Other than their hardiness and long-lasting beauty, all the viticella clematis have the great virtue of not suffering from clematis wilt. This can, however, be devastating for many types of clematis, particularly the large-flowering hybrids from group two, reducing an apparently healthy plant from full flower to brown rags in 24 hours. It is caused by the fungus *Phoma clematidina* entering into a damaged part of the stem. Sometimes only part of the plant will be affected but often the whole thing will collapse. The only course of action is to remove all growth above the point of collapse. However clematis wilt rarely kills the plant and it usually regrows perfectly healthily from a point below the infection. The best defence is to plant deeply with at least 2.5cm (1in) of stem below soil level and to carefully support growing stems so that they aren't damaged by wind.

All clematis grow best when a good amount of bulky organic material such as garden compost or farmyard manure is added to the planting hole. This is as much to retain moisture as to feed the plant as they all grow better with shaded, moist roots.

(Left) **Clematis** *'Madame Julia Correvon' scrambles up a support of hazel bean sticks in the Jewel Garden. It will be pruned back to the ground every March.*

(Following page) **Ligularia przewalskii** *in the Damp Garden creates shimmering yellow spires of flower around pitch black stems.*

LIGULARIAS

If you want a brilliant display of flaming yellow and orange in high summer – and who would not – then you can hardly do better than plant ligularias.

Finding the right spot is the key to good ligularias. We used to grow them in the Jewel Garden where they thrived – until the midday summer sun came out. They reacted to its full glare by flopping and drooping horribly and looking as though they were in the grip of some terrible disease. Cloud or nightfall invariably restored them but the truth is that ligularias must have some shade unless they are planted in boggy ground – by the side of a pool for example.

However, I moved them all to the Damp Garden which not only has moister soil but also high hedges that screen the worst of the sun and wind without casting complete gloom – and they have thrived there ever since.

That was over ten years ago, and this year we completely renewed our stock of plants in this damp area as it had become rather wild, tired and overgrown (which sounds more like the description of an errant rock star than a garden).

I have added three types of ligularia: *Ligularia przewalskii*; *L.* 'The Rocket'; and *L. dentata* 'Desdemona'. Przewalskii is a mouthful for the English-speaking tongue but a joy in any language. It has deeply cut leaves and a 2m (6ft) tall flower spike (on a mature plant) that is almost black and carries a brilliant spire of small yellow flowers from June onwards for weeks. It looks best when threaded through the border like a wonderfully dramatic wildflower in tall grass. 'The Rocket' also has a tall black flower stem carrying a conical torch of small yellow flowers, although these are individually rather larger than its 'Polish' cousin and it has heart-shaped leaves with ragged, serrated edges. (*Przewalskii* is not Polish at all but a native of damp meadows in north-west China.) Both work well together, harmonising on the same theme.

'Desdemona' is also native to China and Japan. I have previously grown *L. d.* 'Othello', which shares the family's superb foliage with green upper surfaces and rich cardinal-purple undersides carried on plum-red stalks that reveal themselves with every breath of wind as it catches and twists the leaves. Their flowers are the colour of orange peel with petals like chunky daisies. Their only fault is that they are slower-growing than taller ligularias and can be prone to slug attack during the first month they emerge in spring, so need extra protection and care at that time.

This kind of planting, taking one kind of flower and then varying that theme with hybrids and species, nearly always looks good and provides an inbuilt harmony to a

border, even if the individual plants vary quite noticeably. This is as much to do with the fact that if one of them is happy in that position then the others are almost certainly bound to thrive there too and in my experience, a happy plant is also a beautiful plant and most certainly leads to a happy gardener.

LILIES

Lilies have become refined and perhaps reduced to one distinct group of flowers yet until fairly recently reference to a lily could cover, amongst others, waterlilies, arum lilies, or lily of the valley. In botanical fact, the lily family includes tulips, erythroniums, fritillaries, kniphofia (red hot pokers) and colchicums (autumn crocus) although does not include lily of the valley, or daylilies (*Hemerocallis*). But in whatever form it comes the word lily suggests a combination of opulent beauty and purity.

There are about 80 species of lily recognised by botanists, with many hundreds of hybrids. The hybrids are divided into different divisions, based upon the dominant species of the parentage. You have Asiatic, martagon, European hybrids, American species hybrids, longiflorum hybrids, trumpets and orientals. Each division is sub-divided further. The species originate from across almost the whole breadth of the Northern Hemisphere, from sea level to over 2,500m. *Lilium regale* is found in only one steep-sided valley in China, whereas *L. martagon* is found in an area ranging from Siberia through Poland and down to the Balkans.

Most are essentially woodland plants and need semi-shade, ideally with their flowers in full sun and roots in shade, and a woodland soil, namely one like leaf mould, rich, well-drained yet moist. We find that lilies in pots – which we grow quite a lot of – always do well if the potting compost is 50:50 sieved leaf mould and normal peat-free potting compost.

They are difficult to store because – although proper bulbs – they must not dry out. But the main problem with lilies comes in the shape of a bright red beetle, *Lilioceris lilii*. These are unmissable and unmistakable and love nothing more than munching on every part of the plant. By far the best way of controlling them is to check each plant every few days and simply remove them by hand.

Lily disease is a fungus that can first appear as dark green spots on the leaves, which will then turn brown and die off. The flower buds wither and fail to develop.

Like most fungal problems it is worse in a wet summer. As a rule, however bad the damage to the plant above ground, the bulb remains unaffected and a new flowering stem will emerge next year. Remove and burn all affected material.

Martagon lilies seem to be resistant to lily disease and we have some in the Jewel Garden that return like old friends year on year, although they took three years to produce flowers after planting. It was worth the wait. Each individual flower is a Turk's-cap type, in that its petals hang straight down before curving right back to form a 'turban' with up to 50 of these hanging from each stem. Mine are a shocking pink with deep plum speckles at their base and orange anthers, but there is a white form, *L. martagon* var. *album*, and a rich maroon one, *L. m.* var. *cattaniae*. They will naturalise in long grass or dappled woodland and are ideal for growing beneath trees or shrubs.

I have *Lilium regale*, regal lilies, in the Walled Garden that are at their best in early July. These were famously discovered by Ernest Wilson in 1903 in a steep Chinese valley where, according to Wilson's own account, 'In summer the heat is terrific, in winter the cold is intense, and at all seasons these valleys are subject to sudden and violent wind storms'.

Wilson brought bulbs home and since then these have been the most popular garden lilies and the easiest to grow. The buds are pinky purple flushed with a touch of grey, sticking out at right angles to the stem like speed boat hulls, and they open white with a yellow flush at the base of the petals. The scent from these voluptuous trumpets is as full-throated as nightingale song. We grow ours in a very sunny corner, in the company of *Acanthus spinosus* and the rose 'Charles de Mills' and for a month in midsummer this is as opulent as an English garden can get: yet the plants take all the wet and cold and wind that this garden absorbs each winter without any trouble at all.

I grow *Lilium henryi* in the Damp Garden amongst the hostas and ligularias and it thrives with practically no attention, producing bright orange flowers from July through to September on stems a good 2m (6ft) tall. It is a truly easy, even friendly, plant to grow, tolerating most soils and having great longevity. *L. henryi* is one of the Asiatic hybrids that contain a huge range and variety including *L. nepalense*, which is also in the Damp Garden but in a more modest clump. However, I am hoping it will spread widely as it has the habit of popping up by virtue of its stolons that creep underground from the mother plant before breaking cover to flower. These flowers are anything but modest with glorious lime-green petals and deep maroon interior.

I guard it jealously because *L. nepalense* is 'family' – it was introduced to Britain by one of my forebears, the plant collector David Don. However it shares the characteristic common to a number of South African bulbs like *Eucomis* and *Galtonia* (although, as its name suggests, it comes from the Himalayas) which is that it likes to be damp in summer – but not boggy – and dry in winter. This is tough to arrange in a low-lying, wet Herefordshire garden.

SUNFLOWERS

As summer moves into its second half I rely on the half-hardy and tender annuals to provide colour and form. The most dramatic of all late-season annuals – and one of my favourites – is the sunflower. The knee-jerk response to sunflowers is the child's drawing of huge plants topped with flowers as big as dinner plates radiating petals like the sun. These are fun to grow, especially for children, and I love them. 'Russian Giant' is an old favourite, guaranteed to grow exceptionally tall, but stake it early and well otherwise it will inevitably crash to the ground sooner or later. Often this will not damage the plants but they very quickly turn to find the sun and start growing with a crick in their neck so that when you right them again they are permanently bent.

When it comes to selecting the colours of sunflowers I find that the velvety rusts, plum, burnt orange and browns that you find in varieties like *Helianthus annuus* 'Prado Red' and 'Velvet Queen' add a velvety richness to a border that exceeds any other plant in late summer.

'Velvet Queen' remains my favourite, but then she was the first that I ever grew. Each flower is not huge but the size of a good saucer and comes in colours ranging from deep crimson through a burnt orange to brown with golden points of

(Left) Each individual floret on a martagon lily's flower stem is extravagantly beautiful and as the plant can carry up to 50 at a time the combined effect is spectacular.

(Following page) The annual sunflower 'Velvet Queen' has exactly the right richness and intensity of colour to set the tone for the late summer Jewel Garden. The flowers first appear in July and continue into autumn.

pollen dotted about the centre. 'Claret' and 'Prado Red' are both remarkably similar, with a plum undertone to their light-sapping colour. In the shade they are darker and can seem to be almost brown. The 'Prado' series has masses of flowers and is practically pollen-free which means that you do not get the staining if you brush against it.

All these dark sunflowers have long been essential components of the Jewel Garden borders and work well with the late summer palette of crocosmias – especially the orange 'Emily Mackenzie' – as well as the large range of marmalade and plum-coloured heleniums, red dahlias, bright orange tithonias and humble calendulas. At the other end of the sunflower spectrum I like the pale varieties such as 'Vanilla Ice' or 'Moonwalker' as well as the perennial sunflower 'Lemon Queen' with its mass of delicate miniature sunflowers.

Sunflowers respond to dead-heading by providing waves of smaller flowers. 'Velvet Queen' is naturally multi-headed, like a great set of candlesticks 2m (6ft) tall, but others that instinctively produce one dominant huge flower will produce extra shoots at the junction of leaf and stem if you keep cutting off the flowerheads. I stop dead-heading from the middle of September so that the birds have something to eat. The finches love them, clinging to their centres and greedily pulling out the seeds from their individual compartments as they store up body fat for winter, and I have seen a crow trying to copy this, balancing with all the grace of a drunkard walking a tightrope.

SWEET PEAS

Although the flowers are beautiful, and can keep on being so right into autumn – it is the scent of sweet peas that I love most. It has unfathomable depths of gentle sensuousness. It is one of those delicate yet persuasive fragrances that the garden produces, along with honeysuckle, roses, wallflowers and violets, that is more enticing and seductive than any man-made perfume and, in my mind, as good as any so-called exotic plant.

Unfortunately the choice of fragrant sweet peas is more limited than the scores of visually sumptuous varieties that are all but scent-free. This is a result of the obsessive breeding and showing of sweet peas in the last 100 years.

The original sweet pea introduced to this country at the end of the 1600s is likely to have been a magenta and purple bicolour now called *Lathyrus odoratus*

'Cupani'. 'Painted Lady' was one of the first bred varieties, being recognised as such in 1726, and it is still one of the best. Very few new varieties were raised until Henry Eckford, the head gardener at Sandywell in Gloucestershire, virtually invented grandiflora sweet peas at the end of the nineteenth century. These have larger flowers and a much wider range of colours but – crucially – retain their fragrance. The plum-coloured sweet pea 'Monty Don' is a grandiflora.

By 1900, Eckford had raised and was showing no less than 115 different varieties. In 1901, one of his varieties 'Prima Donna' sported a flower with greater size and a distinct wave to the upright bit of the flower (the standard). This happened simultaneously in Eckford's gardens and at Unwin's (then a cut-flower nursery outside Cambridge), and at Althorpe, where Earl Spencer's head gardener identified it and showed it as 'Countess Spencer'. The upshot of this coincidence was that the cross between Unwin's wavy sweet pea 'Gladys Unwin' and 'Countess Spencer' produced the Spencer sweet peas. Spencers have long stems and large flowers that make them ideal for showing – and they have dominated the sweet pea market ever since.

The best for a perfect sweet pea fragrance remains the original 'Cupani' or 'Matucana'. Both are bicoloured, as is the pretty pink and white 'Painted Lady'. Other than these, it is best to hunt out the grandifloras that Eckford developed, which will make up most of any so-called 'Old Fashioned Mix'.

We grow the richer colours in the Jewel Garden including 'Purple Prince', 'Black Knight', 'Midnight' and 'Black Diamond' as well as the red 'Gypsy Queen', 'Violet Queen', the bright orange 'Henry Eckford' and the magenta 'Annie B. Gilroy'. In the Walled Garden we have white sweet peas, all with good scent, including 'Dorothy Eckford', 'Royal Wedding' and the ivory 'Cream Southbourne'.

Horticultural convention has it that you should sow sweet peas in autumn but although this will give you earlier flowers, it means storing and protecting the plants over winter so I like to sow mine in early spring. I sow three seeds per 8cm (3in) pot or two per root trainer (all legumes have long roots and need some depth of soil to grow into) and germinate them in the greenhouse before moving them to coldframes

It is combinations that make a border rather than individual flowers. Here, **Dahlia** *'Bishop of Llandaff', sweet pea 'Cupani', opium poppy seedheads and* **Verbena bonariensis***, against a backdrop of golden hop, combine into a natural arrangement.*

to harden off. If one (or even occasionally two) does not germinate I do not replace it. One strong plant will produce as many flowers as half a dozen weak ones, competing for nutrients. However, when you buy sweet peas from a garden centre there might be as many as a dozen in a pot and it is a good idea to thin them out to give them a chance.

Depending where you live (late frosts are the enemy) mid-April is probably a good general rule of thumb for planting out, as long as they have been well and truly hardened off – give them at least a fortnight outside in their pots and be prepared to cover them with fleece if there is a cold snap. They are inclined to sit rather sullenly when first planted out and are susceptible to slugs before they get growing, so keep an eye on them and make sure that they are well watered. Sweet peas respond well to our rich soil but if your soil is at all light or well-drained it is advisable to treat them like climbing beans – that is to dig a pit or trench and fill it with good manure or compost.

They will need tying in every week or so for the first few months but once they get growing strongly will support themselves. Again, conventional wisdom is that you should pinch out the side shoots, as you would do tomatoes, in order to encourage fewer but bigger flowers. I never do this as I am very happy to have smaller but more profuse flowers.

To keep the plants flowering you must keep picking them, because they quickly go to seed as the weather warms up. They are also quick to go to seed if they become dry so a regular soak is essential. We have found through trial and error that if you want the sweet peas to be at their best you need to pick every existing bloom every 8-10 days. As you do so remove any seedpods that appear because the plant will channel more energy into seeds than making new flowers.

VERBENA BONARIENSIS

Whether forming a 2.5m (8ft) candelabra in our Jewel Garden or a graceful single spire in the harshness of the Dry Garden, *Verbena bonariensis* never fails to please. It has fluted stems that have offshoots topped with an inflorescence of luminescent purple florets. The effect is graceful, slender and yet assured. It is the most adaptable of plants, the shimmering violet flowers working perfectly with rich browns and crimsons just as easily as with the yellows, pinks and tawny softness of autumnal colours. It looks good backlit by the sky and it also makes an open fringe at the front of the border through which to see other plants.

It is a perennial from wet, open fields in South America, and when mixed with border grasses it creates the illusion of a monstrous meadow in a border. It is not fully hardy, although in a mild area it will survive winter to emerge in a very bedraggled state. It should then be cut back, ready to regrow. What is likely to defeat it is the combination of cold and wet – which is the norm in winter here – rather than just freezing temperatures.

But it is easy to treat as a biennial, sowing in May to flower the following July and, to confuse the issue, it quite often behaves like an annual, flowering in the autumn of the same year. It performs best in open sunshine, with rich, moist but well-drained soil.

BEETROOT

Beetroot has long been a favourite vegetable and nowadays is heralded as being especially good for you, too. I work on the basis that all fresh, seasonal organically grown vegetables are good for you but certainly beetroot is wonderful stuff.

The leaves are just as important a vegetable as the root. In fact we came to the red swollen roots relatively late, not taking them up with enthusiasm until the 1600s although the leaves had been grown and consumed for centuries before that.

Beetroot will not germinate at temperatures below 7°C (45°F) so I sow two or three seed clusters per medium-sized plug in February and April and perhaps July too to ensure a succession of fresh roots. They germinate in the greenhouse before going out into a coldframe to grow large enough to be planted out. They are ready – as are all plugs – when there is sufficient root to bind the plug of compost in shape when lifted out of the container but not so much that they are rootbound.

(Previous page) Butterflies love **Verbena bonariensis** *and peacocks, like this one, will rise in clouds from the Jewel Garden in midsummer adding another dimension to the display and quality of the garden.*

(Right) Beetroot at the perfect size for harvesting, although they will happily sit in the ground for months and even if they die right back or get too big to eat, the new leaves in spring are very good in a salad.

I prepare their growing ground with a good dressing of garden compost. They respond well to plenty of richness and water and if too dry or unnourished will respond by bolting which will stop root development unless the flowering stem is cut back as soon as it is noticed. When the seedlings have been properly hardened off I plant them out at 23cm (9in) spacing in each direction. There was a time when I thinned beetroot conscientiously but I now like to grow them as a little group of between two and five beets. This is wide enough to get a small hoe in between them and also for the groups to swell out with ease.

The ideal beetroot size is between a golf and cricket ball and we harvest them by pulling the entire batch of plants that grew together in each plug. Growing them in a little cluster also helps stop the occasional over-enthusiastic one getting too big.

The last batch will stay in the ground all winter and although cold weather will reduce some of the roots to hollow, soggy shells, it is astonishing how new foliage can nevertheless appear in spring – a delicious filler for the hungry gap that occurs in spring.

I have tried many varieties but find that 'Bulls Blood' and 'Chioggia' are particularly good. 'Bulls Blood' is one of our oldest varieties, going back over 100 years, and can get a bit woody so it is best to eat them young and tender. 'Chioggia' is, as its name suggests, an Italian variety from the Veneto, which has concentric pale and dark pink rings. It is deliciously sweet but is a notorious bolter so keep it well watered.

COURGETTES

For all their size, pumpkins and squashes can be sensitive things. In a cold summer they will obstinately refuse to grow, sulking in the vegetable plot, surrounded by the yards of bare soil that they need if they are to grow lustily. However, their smaller cousins, courgettes, are generally more robust and will reliably produce delicious fruits within the full range of our thoroughly unpredictable British summers.

Courgettes are in fact a type of summer squash. Their thin skins mean that they do not store, but on the other hand, unlike winter squashes, can be eaten as soon as they are a few inches long. Leave them to grow and they become marrows – although confusingly, a young marrow is not a courgette. A true marrow needs to ripen and when cooked the flesh becomes much more watery and, to my mind, tastes much less good.

Another difference is that the flowers of courgettes are a delicacy. It is almost impossible to buy these so they are one of the (many) advantages of growing them at home. They are delicious dipped in a light batter and deep fried, either just as they are or filled with cream cheese held in place by twirling the end of the flower over.

I sow my courgettes in large plugs or small pots, two to each, removing the weaker of the two plants if both germinate. They need some heat to germinate so a greenhouse or windowsill is good for the early seeds or a propagator that can be put on a windowsill. Prick the seedlings out into a larger pot – at least 8cm (3in) – and harden them off well before planting out after the last risk of frost. All cucurbits thrive in very rich soil with lots of heat and water so add plenty of compost to the soil before planting.

By June the soil should be warm enough to sow directly where they are to grow, especially if you put a homemade cloche (of a cut-off plastic bottle) over them to make a micro greenhouse for the emerging seedling.

This year I have planted my courgettes at the centre of my climbing bean wigwams. This uses the space and should not interfere with the beans as long as I keep each wigwam well watered. I also often grow some courgettes in with sweetcorn. They make a good combination and a good use of available space. The young sweetcorn, planted in a block, do not shade out the courgettes and even when they are fully grown – and by now they are 2.5m (8ft) tall – you can still get between them to cut the courgettes.

The secret of harvesting courgettes is to cut the fruits as soon as they are big enough to handle. It is much better to have half a dozen small ones than one or two big ones: small ones taste better, are no more effort to cook and stimulate the plant to produce more.

Once growing strongly they are pretty hardy and healthy with only the grey, powdery mildew likely to be a problem – and this is always a result of the roots becoming too dry.

(Following pages, left) Deep-fried courgette flowers are a delicious harvest in their own right and collecting them to eat helps reduce the inevitable glut of courgettes that most gardeners experience.

(Following pages, right) Freshly harvested garlic drying in the greenhouse. When all the green of the stems and foliage has wilted the bulbs will be cleaned and tidied for storage.

GARLIC

I always harvest my garlic at some point in July and spread the bulbs out to dry in the greenhouse, which is emptier now that most of the seedlings have moved into the garden or coldframe. The drier the garlic bulbs, the better they will store and I leave them baking until the leaves have completely shrivelled. We then cut the roots off, take the driest outer layer of skin off and trim the tops back before storing them in wicker baskets – although they also keep well and look handsome plaited into ropes that can be hung up to store. But however you choose to keep them they will store best in a very cool, but frost-free, dark place such as a cellar or shed.

I buy fresh seed every other year, keeping exceptionally fat bulbs in the interim for the following year's seed. If you always keep your own cloves for sowing you are liable to build up viruses so fresh seed from a reputable supplier is advisable every few years.

Although garlic grows best with plenty of long hot days in spring and early summer, in order to form a bulb with separate cloves it needs at least a month in the ground when the temperature does not exceed 10°C (50°F). So it follows that the earlier in autumn it is planted the more chance it has of the necessary cool period. Planting between the end of September and the middle of November is ideal. My favourite early varieties are *Allium sativum* 'Thermidrôme', 'Printanor' and 'Germidour' which have large leaves and make large bulbs. However I always plant a second batch of 'Cristo' around the New Year as these later ones, although smaller, tend to keep longer.

Garlic is susceptible to white rot, which has affected the whole of the Ornamental Vegetable Garden at Longmeadow. This is a fungal disease, *Sclerotium cepivorum*, which is most noticeable as a white mould at the base of the plant that rots the roots and thus inhibits and eventually completely stops all growth. The fungal spores of sclerotia – tiny black dots – can remain viable in the soil for at least seven years, meaning that no alliums can be grown at all. So far it has not affected the Top Veg Garden where the garlic produced a good harvest.

Organic growers sometimes spray ground to be planted with garlic with a garlic spray a few weeks before planting. This initially stimulates the sclerotia into growth but they quickly exhaust their food and decline. At this point the garlic is planted when the spores are at their weakest. They grow quickly and can be harvested before the white rot returns to spoil the crop. However, this would only work, I suspect, if the weather was conducive to fast growth in spring.

NEW POTATOES

My birthday on 8 July (just two minutes after midnight) triggers an annual ritual in the garden that is to dig my first new potatoes of the year. I know that there are those who sadly see all vegetable growing as a kind of perverse race and measure their success solely by the date that they can flourish their first harvest, but in this household it is taboo to break the ridges before the end of the first week of July and bad magic to miss it. Some years I can hold the harvest of conker-sized spuds in one hand whereas in others they are swelling into overblown maincrop, a good few weeks past their perfect egg-like size. Regardless of proportion, the potatoes get taken indoors, cooked straightaway and eaten with the luxury and relish of asparagus or the first peas.

And they should be considered a treat that cannot be replicated out of season without losing savour and betraying the whole ethos of such treats. This is right at the core of the pleasure of growing your own. It is the real thing and there is nothing between you and the experience, no one packaging and selling it to you as the 'new potato experience', 'taste sensation' or with the empty promise of year-round satisfaction.

I try to grow first and second earlies and maincrop – although finding space for maincrop is becoming increasingly tricky. But Longmeadow is surrounded by commercial potato growers and their fortnightly spraying means that blight has become endemic. I was talking to an old farmer the other day who said that until 30 years ago they all grew potatoes and hardly ever got blight. 'But it's all different now,' he said with a shake of his head. 'They do it in such a tremendous kind of way.' He was not being complimentary.

In any event, the tremendous scale of the potato growers with their vast machinery and breathtaking investment (and, in a good year, profits to match) means that blight is almost inevitable – especially if you are organic and do not wish to douse your garden with fungicides. There are some maincrop varieties that show reasonable resistance to blight such as 'Sante', 'Sarpa Mira', 'Cara' or 'Orla' – but none are immune, and growing a potato just because you can rather defeats the point. I want to grow potatoes – anything edible – because they are the best tasting, healthiest option possible.

So now, apart from a couple of short rows of 'Sante', I concentrate on first and second earlies. These are varieties bred for their ability to grow and produce tubers quickly. The downside is that they do not store as well as maincrop varieties. Second earlies such as 'Charlotte' fall somewhere between the extremes of a first early and maincrop and will often store well for months.

'Charlotte' is a tried and trusty favourite – perhaps the second best general-purpose boiled potato of them all, which is ready from late July and keeps for months. The best? One romantically named 'BF15' which is almost impossible to get hold of nowadays. Its name derives from sports of 'Belle de Fontenay', and, in the couple of years I grew it, it was almost perfect in terms of taste and texture. Mind you, 'Belle de Fontenay' itself is lovely, too.

The whole point of first and second earlies is not just that you have a supply of spuds to hand, because these are very easily and cheaply bought, but their texture, taste and above all, sweetness. The sugars in a potato fresh from the ground are more intense, and if you cook and eat them as soon as possible after harvesting there really is an appreciable difference from anything that has to be lifted, packed, transported and marketed. New potatoes should be dug specifically for each meal and each harvest cooked in its entirety because leftover cold ones are almost as good as those served steaming hot.

Digging 'Red Duke of York' first early potatoes on my birthday at the beginning of July. Not a great harvest, but 2011 was a difficult year for growing potatoes.

Unlike summer raspberries which need permanent support, autumn raspberries need only strong twine to stop them flopping for a few months – because they are cut hard back at the end of each year.

RASPBERRIES

Raspberries are my favourite soft fruit, infinitely superior to strawberries in every way, and growing them is easy enough – but does need an understanding of how they grow and produce their fruit in order to get the best from them, and it's important to follow a simple pruning regime. But once established they are largely pest- and disease-free, take up very little space, will happily grow in shade and produce fruit that is delicious fresh, freezes well, makes excellent ice cream – and not to forget the best homemade jam that there is.Raspberries will grow on most soils but are happiest in a slightly acidic, sandy loam. They like plenty of moisture, and do not seem to be fussy about shade although I would suggest planting them either against an east or west-facing wall or fence or with some protection from hot sun.

There are two types of raspberry, summer- and autumn-fruiting; and growing both kinds will give you a steady supply from the end of June through to November. The fruits of both are fairly similar, although winter raspberries are, in my opinion, slightly sweeter, tastier and much easier to grow. Some of this ease is due to the weather in late summer and autumn (which tends to be cooler and damper, and suits the fruits), some of it is because the young blackbirds and thrushes that adore raspberries in June and July seem to leave the later crop alone, but most of it is down to the different pruning regimes dictated by fruiting habits. Summer-fruiting raspberries produce their fruits on canes grown the previous summer, while autumn-fruiting ones fruit on canes grown the same year.

For summer fruits I grow 'Malling Jewel', an old-fashioned variety with large fruits and more modest growth, and 'Glen Ample' which has smaller but more

abundant fruits and no prickles on its very vigorous canes, which makes pruning and picking much easier. My autumn raspberries are 'Autumn Bliss' which are delicious. Yellow raspberries taste very good and are less vigorous than most – so good for a small garden. They are nearly always autumnal. 'Golden Everest' and 'Fallgold' are two good varieties.

The time to plant raspberries is between October and March. Prepare the ground well, digging deeply and adding plenty of organic material – although not mushroom compost as this is too alkaline. Set up whatever support system is appropriate before you plant.

The canes should be set in the ground vertically so that the roots are just covered and then cut back to 23cm (9in). This means that there will be little or no fruit in the first year from summer raspberries, but it is an investment that will stand you in good stead as the growing energy will go into the roots and vigorous new growths for the following season's fruit. If you have autumn-fruiting raspberries, the subsequent regrowth from this harsh winter pruning will bear fruit in the first year.

Hoeing raspberries is risky because they have very shallow, fibrous roots. However you can suppress weeds by mulching them very thickly each spring. Composted bracken or pine needles are ideal, garden compost is good, and chipped or shredded wood perfectly adequate, as long as it has been left to compost for six months or more.

RED CURRANTS

Like all currants, red currants belong to the *Ribes* family and have been grown in British gardens since Tudor times. The variety Great Red Dutch, brought back from Holland by John Tradescant in 1611, became very widely popular and for the next couple of hundred years red currants were a much-loved fruit and more popular than strawberries, raspberries or black currants.

They are easy to grow and will tolerate deep shade, wet, cold and almost any soil condition. The best time to plant them – and all fruit bushes or trees – is when they are dormant, between October and February. If left to their own devices red currants will make a sprawling bush, but they can be trained into cordons, goblets, fans or whatever takes your fancy, which makes them very suitable for a small garden.

Although they adapt to most soils and conditions they do best with a mulch of compost in spring. Whilst they should be planted 1.5–2m (5–6ft) apart as bushes, cordons (which are trained as a single stem with fruiting spurs up its length) can be 45cm (18in) apart.

The spurs develop on mature wood so pruning is geared to maintaining a framework of mature wood yet replacing growth every three or four years. This is done by removing the oldest branches every year and encouraging new ones to take their place. I find the ideal shape is a goblet on a stem anything from 15–60cm (6–24in) and an open bowl comprised of four or five branches. To create and maintain this I prune all inward and crossing growth in early March and then shorten the remaining stems by a third. This keeps a good structure whilst encouraging healthy new growth.

Red currants respond well to a feed of potash in spring and I supply this simply by sprinkling ash from our wood fires before mulching the bushes with a generous layer of garden compost (which serves as much to suppress weeds and retain moisture as to feed them). My favourite varieties are 'Laxton's Number One', 'Junifer' and 'Raby Castle'. White currants are effectively albino red currants so should be treated identically. I grow 'Versailles Blanche'.

COMFREY

As a rule, an annual mulch a couple of inches thick of garden compost is sufficient to replace the nutrients used by most plants, and it will improve the soil structure at the same time. But if you grow plants in containers or are raising fruit and vegetables you will occasionally need a little extra feed.

For flowers and fruits the key is potassium and I grow comfrey as a crop for the sole purpose of acting as a potassium-rich feed. This is particularly effective with any plants that produce a large crop from a relatively small space such as tomatoes in a greenhouse.

I grow the wild, self-sown *Symphytum officinale* that you find fringing rivers and all damp ground. It is a beautiful plant with pink and mauve bell-like flowers. It seeds itself easily and can become invasive because it is a big plant, but I treat it as a gift, cutting and pulling the plants I don't want as a valuable addition to the compost heap. If you cut it hard back to the ground, each plant should provide between three and five crops a year.

Comfrey has a long taproot – up to 3m (10ft) in deep soil – and is spectacularly efficient at sucking up available nutrients and storing them in its leaves. These nutrients include trace elements that plants use in minute quantities but which can also be critical to the healthy balance of growth and performance. These break down very fast and so transfer back into the soil exceptionally quickly. Comfrey also has a higher level of protein in its leaf structure than any other member of the vegetable kingdom.

I use the comfrey in three ways. The slowest and most common is simply to add it to the compost heap where the nutrients enrich the whole heap and the plant acts as an activator. It has a high carbon to nitrogen ratio so is ideal for mixing up with grass cuttings.

The second way is to spread the leaves as mulch beneath the tomatoes. This acts like any other mulch, slowing down evaporation of moisture and suppressing weeds but more importantly, enriching the soil right where the surface roots lie. There is a by-product to this mulch which is that as comfrey leaves wilt they become irresistible to slugs and snails so I strew them around young plants like lettuce, that are prone to slug attack, as a delicious distraction whilst my vulnerable plantings get a head start. It works well.

The third and most important use is as a liquid feed that can be used either as a foliar feed or watered directly onto the roots. Making it is easy. I fill a bucket with as many comfrey leaves as it will contain and then top it up with water. The leaves soon turn into a sludge that smells appalling – but a covering of thick polythene will contain it. This stench means that the proteins, which are as high in comfrey as in legumes, are breaking down. After three weeks the mixture will be a greeny brown soup and ready for use. The secret is to keep it well diluted. You do not get better results by increasing the richness. 1:10 is plenty strong enough, and there is never any need to use it more than once a week for plants that are struggling to produce a large crop of fruit or flowers, or once a month as a general boost for permanently containerised plants such as citrus.

The results are not spectacular, nor should they be, but it is very effective. Any healthy plant grows steadily and within the limitations of the circumstances. The trick is to make the most of them.

Because comfrey is such a useful plant and grows so easily at Longmeadow, it is easy to take for granted how handsome it is too.

Natural gardening

One of the primary reasons that I garden is to create a little corner of this planet that I can personally tend and care for. To me this means growing both beautiful and delicious plants but also, just as importantly, looking after the whole mini ecosystem of my garden. A balanced, healthy garden that is self-sustaining and rich with wildlife is just as important as a few prize plants. So for the last 20 years it has seemed pointless to me to garden with chemicals. The less I can interfere with nature, the better.

I am under no illusions however that any garden is 'natural'. Gardening is all about shaping and controlling nature. The real skill, it seems to me, is to harmonise your own horticultural control with the balance of the natural world. It is a relationship. I choose my plants with respect to what I like but also to what will grow well here, in this particular garden. I may have various pests, weeds or other problems but they are almost certainly aspects of my soil, location and choice of plants. In other words they are symptomatic of how I garden rather than trials sent to test me as a gardener. We – every possible living thing in the garden – are all in this together. Even slugs and aphids are all part of the natural balance.

Now, to some degree this is inevitable. Most experienced gardeners learn sooner or later that nature will always win any fight you might choose to pick with her. Go with her, pamper and steer her and problems melt away. Try and impose your will and pride is going to be followed awfully soon by a mighty fall.

This inevitably means encouraging native plants as well as more conventionally horticulturally 'interesting' ones from around the globe. Indigenous plants will always be healthier and quite literally more at home than anything that has travelled across the world to your back garden. But unlike many other parts of the world, British gardeners have developed a tendency to make a distinction between garden plants and 'wildflowers' as though they were automatically in separate categories. Yet has any garden ever been as lovely as a bluebell-carpeted wood, a bank of cow parsley, honeysuckle, wild garlic and meadowsweet, or a close-cropped down jewelled with a carpet of cowslips? I certainly do not think so and try to incorporate the essence of the local countryside into Longmeadow with the same enthusiasm as others might recreate an Alpine rock face or a corner of a tropical rainforest.

Healthy plants make a healthy garden, and to this end it is often better to sacrifice some size or unseasonable harvest of fruit or flower. Plants are healthiest when they are getting all their nutrition from a healthy soil. This inevitably means that they will vary in size or rates of growth. There is no objective standard to this. It is far better to have a healthy plant that is half the size of one that has been forced and fed into unnatural growth. There is nothing that predators or disease like more than lush growth unsupported by a healthy root system.

Over the years I have come to the conclusion that most people over-feed their gardens. It is almost always unnecessary. Your energy is far better spent improving the soil and letting the plant make the most of what it has to offer in its own way. Plants have an amazing ability to adapt and make the most of the conditions that they find themselves in, and a decent soil structure is generally more useful to them than high levels of nutrition.

Rich meadow

A meadow is so much more interesting than a regimented square of lawn – as well as much more ecologically sustainable. However, it is not just enough to let the lawn grow. Do that and it will look fine for a few weeks but grass tends to be a thug and overpower all but the most vigorous plants that grow alongside it.

In an ideal world you must strip the soil of its fertility to weaken the grass and encourage the wildflowers, like poppies, oxeye daisies, cornflowers and what have you, that are adapted to very poor soil. The only practical way to do this is to remove all the topsoil and sow into the well-drained, uncompacted subsoil.

A very good way to get flowers to grow in existing grassland is to use plugs. These are small plants grown in individual pods or plugs and sold by the tray. Each one will be small but have a good root system and will establish and hopefully seed before it can be swamped by grass.

However, this vision of the meadow – a haze of flowers growing in thin soil without competition from overpowering grasses – relates to just one type of flowering mead, typically found on chalk downland. If, like me, you have wet, heavy clay, you are never going to create this type of vision.

There is an alternative and one that I am trying to encourage and establish here in the Writing Garden. This is the wet type of meadow where the grasses cannot be thwarted. Not that I would want that – they are lovely in their own right, especially for about three weeks in mid-June when they are flowering and as beautiful as any more conventional meadow flower. Underneath the grass are creeping buttercup, normal buttercups, red clover, meadow geranium, plantains, sorrel, dandelions and daisies, all wildflowers or weeds according to your inclination. None of it was sown but has grown naturally from the field that we originally made the garden from.

Into this we have been adding fritillaries, narcissi, *Allium sphaerocephalon*, *Geranium pratense* (not called meadow cranesbill for nothing), cow parsley (*Anthriscus sylvestris*), meadow sweet (*Filipendula ulmaria*), devil's bit scabious, cowslips in the open, and primroses along the verges with the hedges. These should all cope with the fundamentally damp, rich soil and more importantly, with the competition from the grasses.

Meadows must be cut, albeit only once or twice a year, and all cut material must be removed and composted so that no extra goodness goes into the soil and encourages grass growth, which will be at the expense of even the most vigorous flower. Obviously when you do this you risk taking away seeds that have not yet dropped, so the timing is critical. In general you should cut it about six weeks after the flowers have faded. In most cases, some time in July is about right – although I shall leave mine till August or even September so as not to cut off the late flowers such as filipendula or scabious. It is likely to need a second cut in autumn or late winter so that it goes into spring short.

Knowing when not to do anything is as great a horticultural skill as any other. The more that I garden the more I take pleasure in making the garden look as natural and unmodified as possible.

AUGUST

The light changes in August and the garden takes on a new suit of clothes. June and July have a growing, swelling momentum but August is tinged with loss. Every day that passes is a little less of summer, the school holidays and the best of the year – the movement is towards the further reaches of autumn. Yet the borders gain an intensity as though to counter this. There is a new richness of colour, especially in the Jewel Garden, dominated by oranges, burgundies, purple and gold.

There used to be a piece of received horticultural wisdom that August was somehow a drab month in the garden and the garden would often be described as 'tired'. People would go on holiday knowing that they were not missing much. Maybe climate change has had an influence but if it ever was true it is certainly not so now. For years we have not taken a summer holiday because to do so would be to miss out on the garden at its very best.

If anything, the August borders take on a kind of muscular energy, full, mature and assured. The sun is lower in the sky and the evenings, by the end of the month, much shorter, so my favourite time of day is when the early evening sun hits the rich colours in the borders so that they glow with regal intensity. Rudbeckias, sunflowers, *Verbena bonariensis*, dahlias, cannas, asters, and the late-flowering clematis all blaze in this low, slanting light. The weather is also kinder for gardeners and plants, with hot sunny days and cooler nights.

Although a skilful gardener can keep a supply of herbs, fruit and veg going for most of the year, August is undoubtedly the best month for the vegetable garden. No other month offers such largesse, combining the last of the earlier crops like garlic, peas and broad beans with the first of the later ones like French beans, sweetcorn, Florence fennel and the sweetest tomatoes of the year. Onions can be lifted and left to dry in the sun and the ground used for brassicas, likewise the peas and beans.

It is a time of change – the warm days and cool nights mean that there is often dew, which helps seeds to germinate fast and makes them less likely to bolt. Next spring's vegetables can be sown and planted just as you gather this summer's harvest. Above all, it is a celebration of the very best vegetables – fresh and seasonal – and grown with love. There is no better definition of good food.

HOT BORDER

Making a good border that lasts all summer long is really a relay race. Each group of plants, from the first spring bulbs through to the last flowers of autumn, does its leg and then hands on the baton. The hand-over has to be smooth and barely perceptible. This is much easier said than done, but much of my own pleasure in flower gardening is managing and tweaking these periods of transition.

The plants of mid- and late summer can never recapture that lovely freshness and vibrancy of May and June, but they can bring a hot, exotic element to the garden and this is the new tone that I try to introduce to the border.

What I look for is an individual plant with both great foliage and fiery flowers. Cannas certainly fit that bill. I particularly like the mixture of dark or striped foliage with brilliant flowers such as 'Wyoming' which has orange flowers, 'Black Knight' which has red flowers or 'Durban' which has chocolate leaves and orange flowers. Cannas like really rich soil that will retain plenty of moisture and are best treated like dahlias. This means either lifting them and overwintering them in a cool, dry box or mulching thickly with soil if you live in a mild area. Each individual flower only lasts a few days but more will be produced from the same flower spike until there are no more buds. It should then be cut back to the next side shoot where a secondary spike will appear. Most cannas will produce three or four spikes by the end of the season.

A much hardier but no less dramatic plant is *Crocosmia* 'Lucifer' with its vermilion herringbone of flowers, and strap-like splay of leaves. The leaves of 'Lucifer' appear like

(Right) For a month or so in July and August **Crocosmia** *'Lucifer' burns as bright as anything else in the garden. Despite its fiery colour it relishes our rather cool, damp soil.*

(Following page) **Angelica gigas** *has the most amazing plum-coloured stem, leaf sheath and flowerhead. It has an irrestistible attraction for insects, too.*

blades from the corms and I start supporting them early so that they do not flop over their neighbours. The flowering stems rise above the leaves to a good 1m (3ft) high, making incredibly fine splays of pleated bud before they open into an upright row of blooms standing on each spray with orange bases and petals of their familiar devilish red. It is as tough as old boots and will take any amount of cold and wet in winter and need no for feeding or watering in summer. After it has flowered – and it has quite a short season – the seedheads earn their place, starting out as a row of green peas flanking the flowering spine and turning ochre into autumn. Towards the end of this month the much smaller *Crocosmia* x *crocosmiiflora* 'Emily McKenzie' will start to flower – and continue till autumn – producing orange blooms against bronze leaves.

You can see by now that I am a huge fan of orange and red flowers at this time of year. They are hot and spicy and exactly the right tone to lift the garden when it might otherwise be getting a little jaded. With careful selection you can have red hot pokers flowering from May through to September but for the late border try *Kniphofia uvaria* 'Nobilis' which is spectacular but needs a lot of space because it is enormous. More modest are *K. rooperi* and *K.* 'Samuel's Sensation' which has dark stems. Both of these will reach about 1.2m (4ft) tall. All kniphofias will thrive in rich, damp soil but will be happy in any well-mulched border.

ANGELICA GIGAS

No plant is more popular with butterflies, hoverflies, bees and wasps than the lovely plum-coloured umbellifer *Angelica gigas*. It starts out slowly, preferring rich, damp soil and easing itself into summer with modest foliage but in late July throws up a 2m (6ft) tall crimson stem topped with a sheafed bud that opens to reveal umbellifer flowers of the deepest burgundy. It is monocarpic which means that it dies once it has set seed, but these seeds will produce a rash of seedlings that can be lifted and moved wherever you wish to place them, so the plant can live on through its offspring for years.

It will also cross breed. We have a few plants in the Grass Borders that have no obvious parentage other than *Angelica gigas* and either angelica or the lesser hogweed. Whatever their lineage they are lovely plants with alizarin stems and white umbels of flower held high.

BUDDLEJAS

The butterflies are at their busiest in August, jostling and flitting from flower to flower like shoppers in a sale to reach the nectar of late-season blooms – and there is no plant in the late summer garden that attracts them like buddleja.

I believe that buddlejas are much undervalued – perhaps because they are so ubiquitous and grow so freely on waste ground. They make superb cut flowers. The scent is heavenly, too.

On the face of it, it is a wonder that buddlejas do not become a weed on the scale of Japanese knotweed or brambles, given that they grow so predictably and prolifically. Surely there is no other plant in the British Isles that grows so readily and profusely in the unlikeliest of places? *Buddleja davidii* in particular will colonise ground that is primarily loose stone like the shingle on the edges of mountain streams in its native Sichuan, in south-west China. They also particularly like lime, hence the predilection for the mortar in brick walls, urban waste ground, stony railway sidings or an untended back yard, and the fact that they relish the thinnest chalky soil in a garden. The seeds are light and winged so are blown large distances in the wind.

Buddleja davidii was introduced to western gardens in 1870 by French missionary Père Jean Pierre David, a keen botanist. He was also the first westerner to observe the giant panda and arranged for one to be taken back to Paris where it promptly died. But *Buddleja davidii* thrived and thrived. In fact, many Asiatic plants adapt very easily to British conditions and this adaptation contributes as much to their widespread status in English gardens as the exercise of horticultural choice.

There are around 100 recognised species of buddleja and although most come from the east, they are also found in North and South America and Africa. Around 100 years before David, *Buddleja globosa* was introduced into Britain from Peru. Another late-spring flowering buddleja that seems to fit the seasonal palette more easily is *B. alternifolia*. Like *B. globosa*, this flowers on the previous year's growth (whereas late-flowering buddlejas such as *B. davidii*, *B. fallowiana* and *B.* x *weyeriana* famously need pruning hard each spring as only the new shoots produce flowers).

But it is *B. davidii* that reigns in all our gardens now with its court of nectar-intoxicated butterflies. Actually the many hybrids of *B. davidii* do not have as much nectar as the species, so are not perhaps so attractive to butterflies, although the white ones like 'Peace', 'White Bouquet' and 'White Cloud' are more nectar-rich than the coloured hybrids.

To the human eye there are really good colours to choose from, albeit on the theme of purple and mauve. The species is a pale purple and earns a perfectly honourable place in almost any garden but the hybrids have been bred primarily to be more intense in colour. We have the blue 'Glasnevin Hybrid', which has quite small, silvery leaves and lavender flowers, 'Black Knight' which is an intense maroon, and 'Royal Red' which is a purple so iridescent red that it is almost magenta, especially when the tiny orange centres to each floret add to the effect.

All buddlejas that flower after June need pruning back hard in March leaving just a couple of strongly growing leaves below the cut. This will stimulate lots of new growth that produces the flowers. You can train an old shrub to have shape by leaving the thick stems and training back from them, like pollarding a willow, but you need to keep rubbing off leaves as they appear on the lower stems.

DAHLIAS

Dahlias are a very important part of the late summer borders at Longmeadow as well as a staple in pots. With regular dead-heading they can flower magnificently for us from July through to the first sharp frost.

Dahlias originate from Mexico and reached Europe via the early conquistadors but in 1566, Spain, looking to hoard its treasures, imposed a ban on all foreign plant-hunting in her new territories and any publication of their natural resources. So, the dahlias gathered from the New World remained in Spain where they were grown as a vegetable but quickly discarded as harmless but tasteless. The flowers were considered a by-product, rather like potato flowers are today.

But eventually a Swede called Anders Dahl got hold of the tubers and bred them and by the time he died in 1789 he had produced some hybrids. In consequence, the plant was named after him, and by the 1830s they had become all

(Following pages, left) **Dahlia** *'David Howard' has orange flowers and bronze foliage that work perfectly within the colour scheme of the Jewel Garden.*

(Following pages, right) **Dahlia** *'Arabian Knight' is darkest in its centre with the petals gradually fading – which gives it a fierce crimson intensity.*

the rage. The dahlia's ability to be interbred into classifiable types appealed to Victorian notions of improvement, and its bright colours and bedding-out habit to their notions of good horticulture.

Because they come from near the Equator they are short-day plants. This means that they respond to heat rather than light to grow and flower and if you keep picking or dead-heading the flowers they will continue blooming right through until the weather becomes too cold. They are not hardy and frost will reduce the flowers and foliage to black shreds and tatters, but it will not damage the plant as long as the underground tubers do not get frozen. Once the tops have been frosted the tubers should be dug up to be stored. In a mild winter this is unnecessary but in the last couple of years anyone storing dahlias in the ground lost most, if not all, of them.

I cut back the dead growth, leaving a few inches of the hollow stems, lift the tubers, removing any soil sticking to them, and put them upside down in cardboard boxes in the greenhouse or potting shed for a week to dry off. It is vital to label them very clearly at this point because if you are anything like me you will inevitably forget which variety is which.

I then store the tubers buried in old potting compost or vermiculite, with the stems sticking out, keeping them in a cool, dark, frost-free corner of the tool shed. It is a mistake to let them get too dry as they will shrivel up but they must not be too wet either as they can get fungal infections and rot. I have found that it is best to pack them into large pots, give them a gentle watering with a fine rose and check them once a month.

In early March I take them all out, check to see that they are healthy and plant them shallowly into 5-litre pots before putting them on a heated bench in the greenhouse. They will respond to the heat by producing fresh shoots from the top of the tuber at the base of last year's stem. When the shoots are a few inches long you can create more plants either by taking cuttings (severing the shoot at the point where it grows) or by splitting the tuber with a sharp knife.

Professionals grow these cuttings in pots for the first year, removing all the flowers so that the energy goes into making big tubers for the following year's show, but I find them very useful as a kind of second rank of flowers – and it is a very cheap way of producing dozens of new plants every year.

When all cuttings have been taken I harden off the mature plants in a coldframe and plant them outside after the last possible frost – as a rule of thumb this is 'after Chelsea' or at the end of May. The cuttings are normally planted in July or even

August and provide a late flush of fresh flowers. They like a rich, well-drained soil and will thank you for watering in a dry spell but are generally unfussy plants to grow.

It is important to dead-head the flowers regularly, cutting right back to the next bud or stem down. This will stimulate new flowers until the temperature nips them in the bud.

Earwigs are a nuisance as they feast on the flowers overnight. There are a number of ways of preventing this. Paraffin poured into the cane supporting the dahlia will prevent the earwigs from resting up there during the day – and they need somewhere dark to go. A matchbox left slightly open, or a small flowerpot, packed with grass, leaves or moss, and placed on top of the cane makes a good earwig refuge and they can then be collected and moved away during the day whilst they sleep. But do not be too harsh with the poor earwigs – they do as much good as harm.

Dahlias at Longmeadow

Jewel Garden: 'Bishop of Llandaff', 'Gypsy Boy', 'Arabian Knight', 'Grenadier', 'Chimborazo' and 'David Howard'

Walled Garden: 'Jescot Julie', 'Sunny Boy' and 'Hillcrest Royal'

We have a wide range of heleniums at Longmeadow, most of which share a lovely warm blend of marmalade and caramel colours.

HELENIUMS

We grow many heleniums at Longmeadow in the Jewel Garden, Grass Borders and Damp Garden. With their rich colours from bright yellow through to brown, carried in generous clumps and drifts, they provide outstanding value in terms of colour and form in the mid- and late-season borders.

Heleniums naturally grow on the banks of rivers in their native North America and when planting them it is worth remembering that, despite being remarkably adaptable, they are most at home in dampish soils with plenty of bright sunshine. They hate light soil in dry shade. They are herbaceous perennials, regrowing with vigour every spring. Although the old growth remains upright and strong throughout winter it must be cut back in March as the new shoots appear.

All helenium flowers share the same characteristic of having a central raised boss, always with a brown ground and often brocaded with yellow like a rich Elizabethan button, around which a ruff of petals is held, sloping down and away like an impossibly gorgeous shuttlecock. The colours of the petals come in almost every variation and combination of yellow and orange, ranging from the pure yellow of 'Goldene Jugend' to the brick red of 'Rubinzwerg'. The most famous helenium of all, 'Moorheim Beauty', is a rich caramel colour with a brown boss.

In fact, heleniums will flower from early summer (like 'Goldfuchs') right through into autumn, with some not beginning their show until well into August. They range in size, too, with short cultivars like 'Red Jewel' reaching 60–90cm (2–3ft) up to an unlabelled whopper we have that's yet to flower that is already 2m (6ft) tall. Height – and flowering time – can be regulated by pinching them out hard or applying the 'Chelsea Chop' in June, making for sturdier, stockier plants that flower a little later.

The best time to plant them is in spring, just as they are coming into growth, and the same applies to dividing or moving them – this should always be done in spring (and not autumn, as the disturbed roots sometimes die back in an exceptionally cold, wet winter). The clumps grow steadily but slowly, so they are never invasive, but every few years they should be lifted and divided by hand. Pull the parent plant apart into as many clumps as it naturally falls into and discard the centre to the compost heap as, like all herbaceous perennials, it is the outside part of the plant that has most vigour. Heleniums also take well from cuttings although it is usually best to do this in June before they come into flower.

Heleniums look tremendous with almost any range of flowers but especially with grasses or as a lower story to the larger daisies such as rudbeckia or inula that flower at this time of year. They are great for wildlife, too. Bees and butterflies love them all summer as they flower and birds will pick over the seedheads in winter.

THE LATE BORDER (ORANGE)

The last weekend in August is the end of the affair. Like so many things in adult life, it is all to do with school. September brings a new term and a new school year and everything changes.

The garden knows not dates nor terms. It has its seasons and distinct rhythms of course, and it is essential to acknowledge these and work with them. But despite this sense of a door closing on the summer holidays, we are now right in the middle of a distinct horticultural season and it is one of the best of them all, not least because it is shot through with a sense of loss that accompanies each passing day.

The light is the very best of the year. How important light is as part of a garden and how little is ever written about it! From the middle of August until October the light has a unique quality: it is still bright and carries with it real warmth but it is also falling in the sky, so there is little of the glare of high summer.

The evenings, in particular, although shortening by the day, have a golden intensity that perfectly catches and sets off the richness of the flowers. The leaves are slowly thinning too, so that shadows have less substance and this beautiful light filters gently through trees, shrubs and hedges like sunlight shining through muslin curtains. I love it.

The plants that thrive in this season tend to come from the Americas. Cannas, rudbeckias, heleniums, sunflowers, dahlias, inulas, cosmos and penstemons all wave an

American flag. There are, of course, wonderful exceptions to this rule, especially from South Africa, like gladioli, leonotis and crocosmia.

In August, Longmeadow – and in particular the Jewel Garden – glows with rich colours. Oranges, purples, burgundies, magenta, copper and gold run like threads through our borders. But the key colour for me is orange. I remember being told that orange was vulgar, but if an orange flower is vulgar then give me vulgar every time – and lots of it.

It is surprising how much orange the garden can absorb. The burnt orange sunflower 'Russian Giant' takes orange to the brink of brown. The heleniums 'Moorheim Beauty' and 'Rotgold' hit their lovely russet stride with shuttlecock flowers, and the annual *Rudbeckia hirta* complements them exactly. *Achillea* 'Feuerland' will start out vermillion and then gradually fade to shades of amber.

Widely grown and rightly so are *Crocosmia* 'Lucifer' with its wonderful fiery flowers and great spears of leaf, and the more modest but longer-lasting 'Emily McKenzie'. Try the lesser known *C.* 'James Coey' to add another intense shade of orange.

I always grow the annuals *Tithonia* 'Torch' which has sumptuously velvet petals and *Leonotis leonurus* with its sprouting rosettes of flowers spaced at intervals up towering stems – which in rich soil and a warm year can reach 3m (10ft) tall. Although when I saw it growing in its native Kirstenbosch, in Cape Town, it was a woody shrub.

Nasturtium (*Tropaeolum*) will swamp a border if you are not careful, especially if you have rich soil, but when controlled it has an incredible freshness. The orange flowers are best produced in really poor soil. Another annual that will be very happy in poor soil is the Californian poppy, eschscholzia, which has probably the most intense orange of any flower on this planet.

This is the high season for dahlias and good orange varieties are 'David Howard' and the rather darker 'Ellen Huston' which both have chocolate leaves. 'Biddenham Sunset' is a semi-cactus type with a sunburst of orange ranging from almost yellow at its centre to almost red in the outer petals, whereas 'Kenora Sunset' is more solidly orange.

Cannas like their soil to be as rich as possible and will repay that nourishment with a display of huge leaves and brilliantly gaudy flowers. 'General Eisenhower' has brilliant orange flowers, 'Wyoming' has vermillion flowers and dark leaves and 'Durban' is another plant that uses the combination of orange flowers and chocolate-coloured foliage.

Bronze fennel, which grows very easily from seed, is a foil to these strong orange hues, along with grasses like the pheasant's tail grass, *Stipa calamagrostis*, or *Carex comans*.

All these plants share two characteristics that make them such a feature of this time of year. The first is that they respond to heat rather than light in order to grow and flower. The other feature they share is a richness of colour. At other times of year – under different light – this can look garish, but in late summer, under the shallower sunlight, this brashness becomes warm and velvety. To get the very best of it plant these late-season flowers so that they face west and catch the evening light in the hour or so before the sun dips below the horizon.

NASTURTIUMS

Nasturtiums, *Tropaeolum majus*, can become destructive in rich soil, swamping their neighbours and even killing established shrubs and hedging if allowed. However, despite this over-fed thuggishness, the flowers are beautiful and last right through to the first frosts and they work perfectly in the Jewel Garden.

The secret of getting the best from them, with maximum flower and minimum leaf, is to starve them of all but the very poorest soil. The conventional advice is to pot them up in the sweepings of the potting shed floor! Although it is easier to regulate their soil in a container – if you grow them outside in the ground the best place for them is either under an established evergreen hedge, which will suck up most of the moisture and nutrients, or within a wigwam of canes or bean sticks so that they can be trained to scramble up. They will also grow in walls and paths where there is no apparent soil.

Although treated as an annual they are in fact perennial and if you have a favourite it will take from cuttings in autumn that can be overwintered and planted out after the last frost. Seeds are best sown in spring and planted outin late May.

They are prone to attack by caterpillars of the large white butterfly and by black aphids and this makes them useful as companion plants for all brassicas as it diverts predation away from your cabbages to themselves.

Echinacea *'Flamethrower' is one of my favourite plants in the garden.*

NICOTIANA SYLVESTRIS

One of my favourite plants of all is at its best in August. This is the giant tobacco plant *Nicotiana sylvestris*. It will, with good soil and a few months growing, reach 2m (6ft) tall with huge, sticky leaves. That makes it dramatic enough. But the flowers are even more stunning, long tubes of white petals hanging off the flowerhead like floral dreadlocks. The flowers will shrink and close in bright sunshine and then open in the cool of the evening, releasing their distinctive, musky, sexy fragrance. The hotter the daytime sun, the stronger the scent will be in the evening because the oils in the plant heat up and then slowly cool in the night air.

They are tricky to buy as plants – I don't know why because I am sure they would sell like hot cakes – but easy to grow from seed. Sow the dust-like seed in March or April in a seed tray and provide some warmth for germination. The seedlings grow slowly and are tiny, but should be pricked out into plugs as soon as they can be handled, and moved into pots to develop a good root system before planting out in early June. They will flower until the first frost.

THE DRY GARDEN

The climate is changing. The trend is towards drier summers and winters and our rainfall is becoming erratic. This does not mean that we are getting very much less of it, just that it is not coming when we necessarily want or expect it. So our gardens are having to become used to longer periods of drought interspersed with sodden wet and even flooding.

This has been predicted for at least a decade now and explains why in 2003 we decided to make the Dry Garden at Longmeadow. The site is a yard where we stored building materials whilst we slowly restored our house. It was unpromising for any kind of garden as there was literally no soil but a layer of tarmac over solid stone. However, the tarmac lifted easily and we dug out about 8cm (3in) of the shale with picks and shovels creating a pair of beds that were like shallow trays about 8 paces long and 2 wide. I got a hose and tried to fill them with water but it drained almost as fast as it went in.

We then barrowed in a mixture of loam (from turves we had lifted and stacked at the end of the garden), leaf mould and garden compost. When you make a dry garden from scratch it is important to add as much goodness as possible to encourage a good root system for plants. If your soil is clay-based it is a good idea to think counter-intuitively and add some horticultural grit as well as compost, as this will open the soil out and enable roots to go deeper in search of available moisture, as baked clay is very hard for plants to get their feeding roots into. In our case the goodness was excellent but the quantity, because of the shallowness of the beds, terribly limited. Not a promising prospect, you might think, for creating beautiful borders.

We also gave ourselves the brief that they had not just low but almost no maintenance. In fact the sparsity of soil has proved a boon in this regard. In the rest of the garden everything – including weeds – grows with astonishing vigour. But in these beds only that which is adapted to the conditions thrives. That, of course, is the secret of making any specialist type of bed – and probably gardening in general. Never fight nature because she always wins. However you can make a garden in literally any situation from a sheer rockface to a bog.

We stocked our dry borders with plants that seemed to be suffering in our normal soil. Sedums in particular are much happier and upright in this kind of ground, but also the giant oat grass, *Stipa gigantea*, which incidentally hates being moved – even to a better home. We would have done better to have bought new ones and started again. There is a white *Cistus* 'Thrive', *Verbena bonariensis*, which has seeded itself everywhere, as has evening primrose, *Acanthus spinosus* and *A. mollis*, irises, rosemary, alliums, tulips, *Achillea* 'Moonlight', mulleins, cerinthe and lavender. Artemisia loves it as do the felted lamb's ears of *Stachys byzantina*. Fennel grows smaller than elsewhere but very happily. Nasturtiums weave through other plants but do not grow lush enough to swamp them. The rose 'Complicata' is doing fine on these meagre rations, and figs, basking against a south-facing wall, are not so leafy as others in the garden but much the most fruitful.

This raggle-taggle, mixed bag of plants (every one scavenged from other parts of the garden) have one common theme other than an ability to cope with little water and that is a much higher ratio of flower (or fruit) to leaf than elsewhere in the garden. Plants that do well in a dry garden tend to have fewer and smaller leaves and to be either more woody or smaller than their cousins on wetter, better soil.

Most weeds hate it. We have never mulched it and simply have a clear out each spring that takes an hour or two at most. That is pretty much it in terms of active gardening although it changes and performs another floral dance almost every day of the year.

Many Mediterranean plants cope with a wet British summer perfectly well, albeit with the same lack of enthusiasm that we all feel, but a wet British winter does for them. There are no such issues with a dry border. I have found that rosemary, thyme, lavender, santolina and the cistus all take the wettest winter months in their stride in this shallow, impoverished but very well-drained soil.

WHITE FLOWERS

Part of getting to know your garden well is having a detailed mind-map of where the light is at any time of day or year and to make sure that some part of the garden is making the most of what is available in every season.

At this stage of the year dusk creeps forward week by week. It is not just the quantity, but the quality of the evening light that has changed too. It is becoming thicker and velvety. This is when white plants look their best, shimmying out from the shadows like a barn owl slipping across the field. White in the middle of the day inevitably seems bleached and consequentially tired and what is intended to be cool, elegant and mysterious simply becomes washed out. White gardens look best at dawn and dusk, so chart the falling light and place them where they can shine out from a soft, falling shade.

White plants can be tricky to place in a border – partly because as white flowers fade they become ugly brown rags that, unlike those of darker colours, must be dead-headed – and partly because of the effect they have on plants around them, either diluting them or acting as spacers, dividing the other colours so that the flow of a border is lost.

White is banned from the Jewel Garden although occasionally cheekily slips under the radar and a pure white poppy will appear from plum or purple heritage. It is always treasured as a welcome maverick and allowed to stay or reverentially picked as a cutflower. However, we grow a selection of familiar white flowers elsewhere, although mainly in the Walled Garden. From the first snowdrops in January, 'White Triumphator' tulips in April, the spring joy of damson, plum and

The flower spike of **Acanthus mollis** *and white sweet pea 'Dorothy Eckford' in the Walled Garden seem paler against the dark green backdrop.*

pear blossom, the many white roses (of which *R. rugosa* 'Alba' and *R.* x *alba* 'Alba Semiplena' are my favourites), white lupins, *Nicotiana sylvestris*, white poppies, many tripods of white sweet peas, *Cosmos* 'Purity', the floating haze of white flowers from *Crambe cordifolia*, galtonias, the flowers of *Hosta* 'Snowden', *Aster umbellatus* and *A. divaricatus*, through to the final autumnal flowering of white snapdragons, white cosmos and 'Iceberg' roses, and the white stems of *Rubus cockburnianus* in mid-winter. But all these apparently white flowers hint at other colours and all need green to seem as white as possible.

The secret is to use and manipulate the range of greens available. White looks best in a garden against green and all the shades between. White flowers tend to have less form and bulk than darker colours – this is because white bleaches out into the space around the plant whereas a rich red or purple creates a clearer volume – so a dark background crisps up the edges and creates volume. It also means that white looks better within defined green shapes and contained areas.

'Silver' foliage makes the area around it seem cool and ghostly and combined with white the whole thing can be deliciously subtle and elegant. It is not silver of course, any more than the white flowers are white, but silvery shades of green and blue. From the palest, chalky greens of artemisia, cardoons, eryngiums, *Stachys byzantina* to the 'blue' hostas like *H. sieboldiana* there is a good range of foliage to bring out the best in white flowers. My own silvery, greeny, blue favourites to set amongst white flowers are *Melianthus major*, cardoon, onopordum, *Eryngium giganteum* ('Miss Wilmott's Ghost') and *Artemisia ludoviciana* 'Valerie Finnis'.

BLACK CURRANTS

Black currants (*Ribes nigrum*) are very different to red and white currants. They do best in rich, moisture-retentive (but not waterlogged) soil and need a thick mulch of manure or compost every spring. They also need lots of sunshine for the new wood and fruit to ripen, so do not grow them in shade. However they crop best with exposure to cold weather in winter and warmer winters may severely limit their cropping.

Unlike red and white currants (and gooseberries) which fruit on spurs that take a few years to develop and which can mature and last for some years, black currants produce their fruit directly from the stem and new growth is the most productive. In fact, in the first year of growth they produce some fruit, lots in the second year and the crop begins to fall off thereafter. So the pruning regime is to cut the oldest stems of each bush right down to the ground, aiming to roughly remove a third of the total, every year, immediately after harvest and certainly no later than September. It is a deeply satisfying job as the wood cuts easily and smells deliciously of black currants as you cut it.

Black currants take easily from hardwood cuttings stuck in the ground in autumn and left for a year before transplanting. Keep any buds on the cutting, to encourage bushy growth. 'Ben Sarek' is a compact variety, good for a pot or small space although you will get a bigger harvest from 'Ben Lomond', 'Boskoop Giant' or 'Ben More'.

Birds love them and they must be netted from the time the berries start to ripen to the last picking – about mid-June till mid-August – or else the entire crop can be stripped overnight. A net loosely draped over canes will do, although a more permanent fruit cage is worth it if you have more than a few bushes. But do not leave the net on over winter – I did once and a light snowfall was enough to completely buckle and ruin the supporting aluminium frame.

The black currant was introduced in the seventeenth century by John Tradescant, the same time as the Great Red Dutch red currant, but it never caught on in the same way and for 200 years was restricted almost entirely to being a medicinal plant, used to treat gallstones, coughs and chest infections.

The Victorians did try and breed bigger berries and there was apparently one variety that had currants as big as gooseberries – although it certainly does not exist today. But it seems that the sole aim was to increase the volume of health-giving juice rather than to develop the flavour. Few plants have a higher level of vitamin C and in the Second World War every schoolchild was issued with a free ration of blackcurrant juice. Even today, 90 per cent of this delicious fruit is used to make juice, with my home county, Herefordshire, the place where the vast majority of all British black currants are grown.

There are few other places in the world where they are grown at all widely. In 1911, black currants (and red currants) were prohibited in the USA because it was thought that they spread a virus that damaged pine trees. In New York state this ban was only lifted as recently as 2003.

(Following page) Freshly dug carrots from the garden may not have the scrubbed uniformity of supermarket ones but taste – and smell – better than anything you can buy.

CARROTS

Everyone grows carrots and everyone should. Fresh out of the ground just swished under a tap they have a rooty, earthy sweetness that is almost completely lost on the tortuous route to a supermarket shelf.

Carrots are biennials that establish a long taproot in their first growing season and umbellifer flowers the following spring. The roots – the bit that we eat – are only edible before the flower stems begin to form. They will grow anywhere that there is reasonable drainage in a soil free from stones although they like sandy or limey soil best, so our heavy clay is not ideal.

They are in the same family as parsnips, parsley and celery and all should be grown in a rotation to follow the brassicas. The ground should be well-dug and rich in well-rotted organic material (such as garden compost) but not have had any manure or compost added within the previous year as this will cause splitting.

Carrots can be sown at any time between early spring and midsummer as long as the soil is warm. Sow very thinly, either in shallow drills or broadcast across the surface, and then rake in. I now always prefer the latter, harvesting a clump at a time made up of roots of different sizes.

Carrots are naturally not orange but purple. It was not until the sixteenth century that orange ones were bred and it took about 100 years for these to become the norm. Nowadays you can have yellow, red and purple carrots – and every shade of orange. They can have round or stumpy roots, pointed ones, blunt ones, slim ones and great long fat monsters.

The earliest crops come from Amsterdam-type cultivars that have narrow, cylindrical roots with smooth skins. 'Nantes' types are generally bigger and can be grown both as early and maincrop. 'Chantenay', 'Berlicum' and 'Autumn King' types are all best for maincrops that you would expect to harvest in autumn and can have the biggest roots.

Carrot fly, whose larvae bore holes in the roots, is the biggest troublemaker and is best avoided by only pulling or thinning in the evening and covering the growing carrots with fleece. A low hedge, either temporary of an aromatic plant such as chives or permanent like our low box hedges in the Ornamental Vegetable Garden, creates a barrier that the low-flying insect will not go over. Alternatively, covering the ground with fleece from seed to harvest will also be a carrot-fly proof barrier.

CUCUMBERS

For the past few years I have grown cucumbers, melons and aubergines in the propagating section of the bottom greenhouse (which is the one nearest the house). This unheated inner section of the greenhouse has a heated bench with a mist propagator so it can get very warm and humid. Ideal for seeds, cuttings – and cucumbers. The only downside to this arrangement is that there is not much room, however I manage to cram in half a dozen pots of cucumbers, growing the plants up tripods made of bamboo. Ideally the pots would be bigger but it is a trade-off between available space and perfect growing conditions and the only negative is that the plants exhaust the compost earlier than they might.

Above all, cucumbers need warmth and wet to thrive. If you want or have to grow them outdoors it is best to grow ridge types, which are much hardier and have tougher, rougher skins, such as 'Bush Champion', 'Crispy Salad' or 'Marketmore'. I have also grown 'Crystal Apple' cucumbers – which are small and round – in the vegetable garden without any trouble other than from slugs and snails, which love the texture and taste of their stems. Indoor varieties tend to have smoother skins. I have grown cucumbers very successfully outside, but they must have 3–4 frost-free months from seed to maturity and we often get frosts here as late as mid-May – and June can be chilly.

I sow the seed in late April putting them on the heated bench to germinate. Once established I pot on the seedlings to develop a decent root system before putting them into their final container, which should be as large as you have space for. Certainly any container smaller than 10 litres is inadequate. I add a generous amount of garden compost to the final potting mix which helps feed them as well as holding plenty of moisture. If they are to be planted outside it is best to dig pits or a trench and add lots of compost.

Once the fruit start to set and grow it is important to pick them often as this stimulates more fruit. It is also much easier to cope with a few smaller cucumbers at a time rather than a glut of whoppers. A large fruit can be cut in half whilst still on the vine, the cut end will callous over and the rest can be harvested later.

To ensure straight growth they need to hang cleanly from the vine. One way to ensure this is to train them up canes set at a 45-degree angle. The growing fruit will then hang vertically, unimpeded by surrounding growth.

Traditionally cucumbers produced male and female flowers. If the male flowers pollinated the female ones the resulting fruits were bitter, so the male ones (which could be identified by the lack of embryonic cucumber swelling behind the base of the flower) had to be pinched off on a daily basis. Nowadays it is easy to get all-female flowering varieties such as 'Helena', 'Carmen' and 'Media' that produce uniformly sweet offspring. However, if stressed – particularly by cold – the plants may still produce male flowers and these are removed to avoid bitter-tasting fruits.

Cucumbers grown indoors are susceptible to red spider mite, cucumber mosaic virus and powdery mildew. I have never had red spider mite in my greenhouses, which tends to build up when it is very dry. Damping the floor down and watering regularly helps a lot. But I have had trouble with powdery mildew and good ventilation is the best prevention. Cucumber mosaic virus is carried by aphids which will yellow and stunt leaves.

(Following pages, left) Most cucumbers sold nowadays produce female-only flowers, but any male flowers (identifiable by the lack of swelling behind the base) must be removed.

(Following pages, right) Cucumbers thrive best in moist heat so I grow them in large pots in the propagating greenhouse.

The secret of growing good Florence fennel is to plant into rich soil and keep it unchecked by cold or lack of water.

FLORENCE FENNEL

Florence fennel (*Finnocchio*) has evolved – with much help from breeders – from the ordinary herb fennel to a vegetable. The base is swollen and forms an overlapping succession of layers, like a bulb. It is delicious, either raw or cooked, stores quite well and is, to me, an essential taste of late summer and early autumn.

The problem with growing lies in its sensitivity to drought. Any check at any stage in its growth seems to set it bolting, especially if you sow before midsummer, as it is very sensitive to changing day length. The plant develops a hard core – rather like a bolting leek – leaving just the outer layers of the base as edible. Ideally one should try and grow it as fast as possible so really rich soil and a continuous supply of water are essential.

I grow it in plugs, in the greenhouse, which I then pot on into a rich compost mix in 8cm (3in) pots. I leave these in the coldframe to mature and harden off before planting out in rows at 23cm (9in) spacing when each plant has developed a root system. This is usually at the end of July although this year I had the fennel in the ground by early June and sure enough, they all bolted. I make a second sowing as late as early August for harvesting from September until the first hard frost rots them.

PLUMS

I have an assortment of plums here at Longmeadow including dessert plums, a nameless cooker and a bullace in the Walled Garden, which were all here when we came, plus one hitherto fruitless gage and four or five damsons. (It is not so much a matter of arithmetic as classification: there is much debate as to whether 'Shropshire Prune' is a damson or a plum.) But I confess that, other than planting, I hardly tend them in any way at all.

Negligence is usually exactly the right course of action with any kind of plum. I get scores of letters asking how to treat silverleaf, leaf blister, bud drop and the other various ailments that the plum family get, and the answer is always the same: doing nothing is always better than doing the wrong thing or even the right thing at the wrong time.

Pruning is the one area where you need to treat plums right. The first simple rule is, if in any doubt, do not prune at all. Leaving any kind of plum unpruned will not do much, if any, harm. Pruning one at the wrong time of year, on the other hand, can certainly cause problems and even kill it. Like all stone fruits they should only be pruned in their growing season, namely from April to August. If you prune outside this period you risk exposing the plant to infection by silverleaf and bacterial canker, both of which can be serious. Prune growing plants to train them as you wish – and plums train very easily as fans against a wall or fence – but once established as a tree only cut them to remove any crossing or dead growth. Plums fruit on one- and two-year-old wood so the regrowth from pruning will not produce any fruit for a couple of years.

Plums – or at least damsons – are part of the landscape in this countryside along the Welsh border. Almost every hedgerow has a few scruffy little damson trees in it and our boundary hedge is no exception. The local town, Leominster, used to have a factory dyeing gloves and used damsons as the source of a particularly rich red colour – and anyone who has ever squashed a damson on a prize shirt or jacket will know just how permanently it can stain.

Harvesting plums. The fruit of this particular tree hanging over the Dry Garden are very plain eaten raw but make delicious jam.

I planted the plum variety 'Czar' (good for a north wall and pies), 'Victoria' (good all-rounder if a little bland when raw) and the two gages 'Oullins Gage' and Old English gage. There were a couple of damsons in the hedges to which I have added 'Merryweather Damson', 'Farleigh Damson' and 'Prune Damson' – although the latter is also classed as a plum. It is certainly very plum-like and, unlike most damsons, regularly makes good eating straight from the tree.

Our native plum is the sloe, which is the fruit of the blackthorn, *Prunus spinosa,* and the sweet or domestic plum is thought to be an offspring of the cherry plum and the sloe.

All plums do well in slightly heavy, damp soil although perhaps it is better to say that they do not do so well on very hot, dry, free-draining sites. Quite a few varieties, such as 'Victoria', 'Czar' and most of the greengages do well in some shade, even on a north-facing wall. When grown in shade they develop their sugars more slowly and consequentially have a richer, more complicated flavour.

Although I suspect I am out of kilter with most people in this, in my opinion most plums – other than a ripe gage – are improved by cooking. Plum crumble is a treat, plum jam brings a hit of sugary summer into the middle of darkest winter and greengage jam takes this to another level. Plain stewed plums, with a generous dollop of yoghurt, are as good a way to start an August day as any.

I have one scruffy, unidentified tree in my garden that has borne great clusters of completely insipid fruit from the day we moved here which, when stewed, become superbly sweet with an aromatic, lingering aftertaste. It is a miraculous transformation.

Even with dessert plums – those specifically grown to be eaten raw – you often have a glut of fruit all ripening at once and cooking is the best way of not wasting this harvest.

The purple climbing bean 'Blauhilde' whose beautiful, stringless smooth pods are extremely good to eat. Like all beans it grows best with plenty of moisture and heat.

Growing climbing beans up extra tall bean sticks might make picking a bit more complicated but they look magnificent.

TENDER BEANS

Broad beans, dwarf beans, climbing beans, runner beans, kidney beans, beans whose pods are slim bootlaces and beans that bulge and rattle in the husky pod – they are all good. Of course we eat nearly all these fresh and green nowadays but most legumes used to be grown for drying and storing with the consumption of them fresh and green reserved for a treat.

The treat is still there. Beans of any kind straight from the garden have a delicious summer freshness, are incredibly good for you and, as if that was not enough, very good for your soil, too. This is because all beans are legumes and all legumes have the ability to take nitrogen from the air and transfer it to the soil so, unlike most of humankind, they leave the earth a better place than they found it.

By mid-July, the broad bean season comes to an end – although in a cold summer like 2011 that can stretch into August. They overlap and are soon replaced by French beans in this garden.

Sow them too early and the weaklings will be eaten by slugs before they flower, but a sowing of dwarf beans in July will respond to warm summer nights by growing really fast and give a good crop in September and October. Bush beans range from the skinny Kenya-type whose pods are eaten tiny to the ones grown for their shelled beans, like pinto beans. I have grown a wide selection and keep coming back to the yellow and purple varieties like 'Golden Sands' and 'Annabel'. Incidentally, purple dwarf beans can withstand colder weather than most other types.

I love yellow dwarf beans although my favourite is a climber that will grow 3.5m (12ft) or more called 'Burro d'Ingenoli'. It has a curiously buttery texture when cooked, pod and all, and is delicious, especially as part of a ratatouille. This will soldier on well into autumn, as, of course, will runner beans.

Four types of runner bean were introduced in 1633 and two of them, 'Painted Lady' and 'Scarlet Runner', are still being grown in thousands of gardens and allotments today. Although we grow runner beans as annuals, they are a perennial occurring naturally at around 2000m in Central America, growing best on the north side of the mountains. This means that they like shade and lots of water – and so are ideally suited to an average British summer, but cannot cope with drought. Traditionally they were sown in trenches in open ground that had been filled with rough compost and paper to soak up and retain water.

In America, runner beans are grown as decorative plants only and most countries grow them for their seeds and regard our love of the sliced pod as an example of British eccentricity. A final twist – whereas all other climbing beans twist anti-clockwise as they grow, runner beans grow stubbornly clockwise. No wonder we British love them.

TOMATOES

For anyone who only knows tomatoes through the cold, thick-skinned, bland-tasting supermarket imposter of a fruit, then the joys of the real thing await you. Tomatoes picked properly ripe and warm from the sun smell musty and fruity and have a succulent, intense sweetness. My own feeling is that cold, unripe tomatoes are not worth eating. Better to enjoy them in season from July through to November and then eat them as a sauce or chutney made from the excess fruit of summer.

Tomatoes, like all plants that come from near the Equator, respond to heat rather than light to stimulate growth. Assuming that any heating will be minimal (and increasingly expensive) even greenhouse tomatoes will stop developing new fruit by October and the green fruit will be slow to ripen – although I have seen tomatoes ripening at Christmas in a greenhouse in Guernsey. So to have tomatoes by the middle of July you need to sow the seed at the beginning of March and provide them with protection from cold. They need a soil temperature of at least 15°C (59°F) to germinate and the air temperature should not drop below 10°C (50°F).

Fruit of 'Costoluto Fiorentino' displaying their tendency to extreme ribbing – and also the poor setting of fruits in what was a particularly cold summer.

I sow my seeds thinly in a seed tray at the end of February on a heated mat in the greenhouse. I used to sow as early as the New Year but found that I would have scores of plants in pots by early April cluttering the greenhouse and coldframes, ready to be planted into their final growing positions but the chance of frost too great to risk that exposure. I have found that it is better to delay proceedings a little and to have steady temperatures rather than risk the hot days and cold nights of May. Outdoor tomatoes can be a month later and should not be planted until the nights are reliably warm – whenever that might be.

Once the seedlings have germinated and there are two true leaves, the seedlings should be pricked out into small pots or large plugs filled with rich compost. I prefer 8cm (3in) pots. Keep them under cover and water at least once daily. They will grow quickly and be ready for planting out when 15cm (6in) high.

Tomatoes are tough, adaptable plants and will grow in soil, compost, pots, bags, bins or whatever you have. I prefer to grow tomatoes in soil if possible and the raised beds in the top greenhouse are ideal. I also long to grow all my tomatoes outside. I do grow some every year but in truth only one year in ten is a success. Last year most of our outdoor tomatoes got blight but I completely defoliated mine and they survived, came back and gave a harvest, albeit a pretty poor one.

I used to grow 100 plants a year, in a tunnel that stood where the Top Veg Garden now is, to make into a sauce that could be frozen and last all year. It was our staple food. Now we make do with half that number crammed into the much smaller greenhouse. An outdoor harvest would open the doors to year-round tomato heaven once again.

As I explained on page 126, tomatoes can be successfully grown in a container of any kind as long as there is reasonable room for the roots to spread. For many years I always grew some in conventional growbags and I still put any extra seedlings into 20-litre pots. The resulting fruit are of a high quality although the quantity is limited. It is worth bearing in mind when you plant them that tomatoes will need quite substantial support, even when in a container, and a single cane stuck into the pot is unlikely to suffice. They will also need extra feeding, especially as the fruits are setting.

What is necessary is plenty of water, especially for tomatoes in containers under cover. Tomatoes are tough but do best with a steady water supply just as they do with steady heat rather than dramatic variations. Once the fruit start to ripen, however, reduce the water a little otherwise the skins have a tendency to split.

Other than a very few varieties, such as 'Roma', which do better as bushes, it is best to grow tomatoes as cordons. This means tying the main stem to a string or cane and pinching out the side shoots that grow between that stem and leaves. These grow at 45 degrees with great vigour and by removing them you let light and air in and focus the plant's energies into maximum productivity.

When the cordon reaches the top of its support I cut off the top of the plant to stop any new fruits forming so that the existing unripe ones will ripen faster. Inevitably you will have some green fruits but none of these need to be wasted. Store some in a dark drawer and they will ripen, and use others for green tomato chutney.

Tomato problems: Tomatoes are tough plants that can take a wide range of situations and problems in their stride as long as they are warm and have a steady water supply. Big fluctuations, which are normal here in May and June, will stress the plants and expose them to problems as a result. Most problems are avoided by ensuring good ventilation, especially at the base of the plants. But it is important not to strive only for unblemished, perfectly formed fruits. The really important thing is taste and other than sunshine, over which we have no control, the choice of variety and all cultivation should be ultimately geared towards what happens in the mouth. (For more on tomato problems also see page 126)

Storing tomatoes: We store our tomatoes by cooking and freezing them. The easiest way to do this is to cut them in half and lay them on a baking tray with a sprinkle of olive oil and perhaps a few twigs of thyme. Bake them in a hot oven for 30–40 minutes. Either bag them as they are or whizz them up in a mixer before bagging them for freezing. When you come to use them you can cook them as they are, as well as adding oil, herbs, onion and garlic for a delicious tomato sauce.

WINTER SALADS

What cheers me up throughout winter is a regular supply of fresh salad leaves. In fact, the simple process of gathering material for a daily salad becomes both a reaffirmation of growth and life and a piece of counter-seasonal bloody-mindedness. The time to prepare that midwinter harvest is at the height of the summer holidays, in August.

The key to this is the declining light. Although it is hot, the days are drawing in and seedlings, which respond to light more than heat, are growing more slowly. Anything sown much after the beginning of September will struggle to grow enough to give you a harvest before February or March. I start in the first fortnight of August with slow crops like parsley, mibuna, mizuna, corn salad and endive, as well as an autumn-cropping batch of quicker leaves like rocket and almost any lettuce you like to eat. In fact, these late lettuces are often amongst the best of the year, relishing the cooler nighttime temperatures and warmth of the soil.

After this first sowing – which may all get used by late autumn or Christmas – I do another in late August, and another in mid-September, adding in lettuce varieties that are specifically adapted for low light levels and low temperatures such as 'All The Year Round', 'Chicon De Charentes', 'Merveille de Quatre Saisons', 'Winter Density' and 'Valdor'.

Whilst I always start these under cover in the greenhouse, carefully pricking seedlings into individual plugs and then hardening them off in coldframes, it is important to expose them to life in the great outdoors as soon as possible, looking to protect them from midday sun more than nighttime chill.

At this point, I have tough choices to make. In the past I have clung to my tomatoes too long so that carefully nurtured salad crops wait too long in their plugs for space to be freed up. But I have to clear all the tomatoes no later than mid-October rather than hanging on to gather a few more ripe fruits. I then dig the ground over, lightly work in an inch of compost to boost the soil and plant up the winter salads.

Mibuna and mizuna need much more space than you might think if they are to develop large enough to last a winter. Mizuna, which has sharply serrated leaves, tends to be more hardy but the strap-like foliage of mibuna is delicious and with some protection – and if spaced widely enough apart – will make very large plants that can be continuously picked.

Likewise rocket. I am looking for strong plants with big root systems that will withstand repeated gathering of leaves, whereas in spring I now broadcast most of my salad leaves and harvest young on the cut-and-come-again principle.

When it gets very cold, I cover them up with fleece, shut the doors tight and they lie snug and hunkered down until the cold passes.

Hedge cutting

I like hedges very much and have a lot of them both in length and height. Nothing adds such a good combination of structure, shelter for humans, birds, mammals and insects as well as plants, or leafy volume to a garden. Hedges define spaces and I strongly believe that spaces, rather than their contents, make for a good garden.

Good hedges need regular trimming and if they are good tall hedges, such as we have here at Longmeadow, then that can be quite a performance with trestles, ladders and an awful lot of trimmings to clear up.

But it is certainly worth it. If I had to choose – heaven forbid – between losing my borders or my hedges I would sacrifice the borders and all their contents and just keep the hedges with grass between them and know that I would still have a beautiful garden.

When you choose to cut your hedge will have a profound effect on its growth. This is because you are pruning each hedging plant individually and the activity of when and how to prune – and this applies to any plant – will either stimulate or restrict growth.

A hedge left uncut will soon become a row of individual trees with all lower growth suppressed. But if a hedge is clipped regularly it will respond by becoming increasingly twiggy and dense, and the regrowth will be a mass of small shoots until one of them dominates and grows away and stifles the growth of its fellows.

Deciduous hedges are best cut twice a year, preferably once when dormant – between October and March – to establish crispness and to stimulate vigorous regrowth, and once in the growing season to keep that new growth looking trim and

Clearing up can take as long as the hedge cutting itself but in midsummer all hedge trimmings are soft enough to be mown or shredded and added to the compost heap.

sharp-edged. Here at Longmeadow we prefer to cut our hedges in August and January although we also give the occasional light trim in early June and October, especially to what we call 'windows and doors' – all those vertical planes that lead by foot or eye from one part of the garden to the other.

Whatever you use to prune, always make sure it is sharp and regularly sharpen it. This is much safer, better for the plant and makes it much easier to do.

The base of a hedge should always be wider than the top. If you cut the sides dead straight then the top of the hedge will shade out the bottom. A slope or 'batter' lets light get at the bottom half of the hedge, which in turn means that it maintains its thickness and density right to the ground.

All hedges, but in particular hornbeam, grow less well if shaded by other plants. The more light and air a hedge has the thicker and better it will grow.

Evergreen hedges like yew and holly are best pruned in August although more vigorous hedges, like Leylandii, box and privet, will need at least one more trim in spring. Never prune an evergreen hedge in winter as this makes it prone to frost damage in very harsh weather.

Other than hawthorn (which should be reduced to half its height immediately after planting) do not trim the top of the hedge for the first three years and then just reduce the growing tip enough to level the young hedge until it reaches the desired height. But cut the sides back hard each year, keeping the growing hedge narrow, to encourage bushy lateral growth.

An overgrown deciduous hedge can be reduced dramatically with a saw and will regrow with renewed vigour and midwinter is the ideal time to do this. A thick mulch around the roots spreading at least 30cm (1ft) either side of the hedge will give it a boost as it regrows in spring and help suppress weeds.

We shred or mow all summer hedge trimmings and add them to the compost heap and they very quickly make excellent compost. Winter hedge trimmings are shredded and used for mulching trees and shrubs – and young hedges.

Root pruning in August

You do not have to prune just by cutting the top growth. There is as much growth underground as above – although much of it is in the form of tiny hair-like roots. However, every millimetre of this root is vital and influences the life of the plant above ground. Cut or restrict these roots in any way and the hormone activity will alter and the plant's growth rate and pattern will respond. Roots grow most vigorously between midsummer and mid-autumn, storing food for the plant to use in its regrowth the following spring. When it comes back into leaf in spring the storage shifts to the leaves and stems where it is used, with any excess going back to the roots.

If you root prune in summer – which is effectively what you do if you repot or move a plant whilst it is still growing – you will not affect the balance of stored food between root and leaf but will inevitably alter the balance between water taken up by the roots and lost through the leaves by transpiration. So whenever you move a growing plant it is always a good idea to reduce the top growth or take off some foliage and also to mollycoddle the plant with extra protection, water and perhaps feed to encourage fast root growth. The worst thing that you could do is to give it extra nitrogen which will cause a mass of foliage – because there will not be the corresponding root growth to support it. This will result in weak growth that will attract aphids and disease and a cycle of plant sickness that might well kill it.

Because of active root growth at this time of year, quite large plants can be moved now even though they are in full leaf or flower. But you must cut them back hard after replanting and let the roots grow well before they become dormant in autumn. Then the following spring they will be better established in their new home than if they had been moved in winter.

SEPTEMBER

For most of my life I have found that September, although obviously beautiful, has filled me with dread. It was associated with going back to school, with the end of summer and the approach of a long, dark winter and the gradual loss of all things that I love most about life. But I have learnt to love it. I love the slight chill in the air and the way that the garden is spilling with fruit and seed. If the year is slipping away, at least it is showering us with gifts.

The Flower Garden is wonderful in September not least because the combination of emaciation and low light means that you can look through everything. From April to the end of August the entire natural world is building layer upon layer on top of itself and really good garden design uses that to obscure and reveal plants and views so that at every turn you are met with the unexpected – or at least change.

But now the garden is thinning. It is not so dramatic as the leaf fall that is to come, but everything is reducing. You can see through hedges that just a few weeks ago were an impenetrable barrier and light spangles through branches and leaves that formed an umbrella of shade in midsummer. The process of reduction will continue for a while yet, and the light will steadily get lower and lower, but it is only for a few more weeks that there is enough left in the garden to give real interest to the picture.

September may be misty and mellow but it is also the month to celebrate bright and brilliant colours. This slipping of the light is a great advantage to the rich colours in the Jewel Garden as the lower slant of the sun burnishes all the reds, purples, coppers and oranges. All the plants that originate from close to the Equator, like sunflowers, dahlias, cannas, tithonias or cosmos, continue to fill the borders with splashes of bright, strong colour (especially if you can keep up the increasingly

demanding job of dead-heading). Asters and heleniums really come into their own and the late-flowering clematis is often at its best.

The Ornamental Vegetable Garden is still producing many summer crops such as beans, lettuce and sweetcorn. The autumnal harvest that includes brassicas, squashes and celery adds fresh flavours along with the pears and early apples. It is a rich month in every part of the garden, made more precious by the daily awareness that it is all gently fading into autumn.

ONOPORDUM

There are gardeners who regard the giant cotton thistle, *Onopordum acanthium*, as a weed, but I love it and carefully dig up its seedlings where they pop up in clusters and redistribute them for better effect – usually to the back of the Grass Borders and the Walled Garden. I would grow more if we had space.

It is certainly intrusive, growing to at least 2m (6ft) – and often half as much again – with a spread of 1.2–1.5m (4–5ft). But it is amongst the most magnificent plants that any garden can grow with huge grey leaves coated with a milky down, fringed with wicked spikes and great candelabras of flowering stems. It needs strong staking or else it will inevitably be brought crashing down like a felled tree in the first summer gale.

When the plant has flowered it rapidly becomes a spectacular skeleton which, if supported, looks majestic right into winter. The lower leaves must be carefully removed as they fade or else weeding around them becomes a nightmare as the spikes get spikier with age.

Once it has seeded each plant dies but dozens of seedlings will appear around it the following spring. These will spend their first growing season developing a rosette of leaves above ground and a very long taproot below. It is important to dig up all but one or two before the taproot develops – by late spring – and transplant them with generous spacing. In the second spring the rosette very rapidly becomes the familiar tower of silvery thistle.

The familiar Scotch thistle flowers of onopordum can be carried as high as 3–4m (10–12ft) in the air by the time they open.

LILIUM HENRYI

No bulb gives such extraordinary value as the orange Turk's-cap lily, *Lilium henryi*. We planted a few dozen bulbs in the Damp Garden several years ago and expected the display to last perhaps three weeks in high summer. We got the display all right – the stems carrying the spotted orange flowers all reach 2m (6ft) tall or more in our lush soil – but the extraordinary thing is how long they last, flowering for many weeks well into late summer.

The foliage is a deep glossy green and more spear-like lower down the stem but becoming shorter and broader as it nears the top. It is stem-rooting so should be mulched each year with good compost to provide extra goodness for new roots. It comes from China and grows best in chalky soil but as long as the ground is not too acidic it is remarkably tolerant and seems to relish our heavy clay – although we do add lots of compost, and any organic matter will improve the soil structure. It will come back year after year and needs no attention whatsoever. We do not even stake it as the neighbouring plants prop it up. A superb late summer plant.

LEONOTIS LEONURUS

Leonotis has long been one of the mainstays of our late-summer, autumnal borders with its towering fluted, almost square stems and coronets of intense, rich orange flowers that are arranged in tiers all the way up the stem.

It is one of the latest annuals in the northern hemisphere, not performing until late August or September, but it does add a tangerine blaze of colour right into November. In its home in the Cape in South Africa, it is a very woody and substantial Fynbos shrub but a sharp frost will kill it stone dead. It grows very easily from seed, so we treat it as an annual that we sow in mid-May. We grow it under the protection of the greenhouse, harden it off gradually and then plant it into its final positions in early July (after we have created gaps in the border by cutting back early summer perennials). You can also take cuttings in September from side shoots and overwinter these to plant out the following spring.

In 2011 it was too cold for these young plants ever to get growing but in a reasonably warm year it will reach up to 3m (10ft) tall in good soil and yet is always slender and elegant and can be fitted into small spaces amongst neighbouring plants. Although never becoming the woody shrub that it wants to be back home, by October the stems are very tough and need shredding to compost well.

LATE SALVIAS

It only needs the first frost – which often comes in September here on the Welsh Borders – to make me realise how precious the last few flowers of autumn are. Of this small band the salvias are shining their light brightest.

I say 'the salvias' as though I have a large collection of them. I do not. But I do have three different flowering types. My favourite is *Salvia guaranitica*. I have *S. g.* 'Blue Enigma' and the flowers are blue until you put them next to true blue when they are shown to be a deep and lovely violet. The flower tube of 'Blue Enigma' is a gleaming satin but the lips are covered in velvety down. They start flowering in June in a warm year although it can be as late as August if we have a cool spring and early summer. I was given my original plant as a cutting 20 years ago and although I must have taken hundreds of cuttings from it since, the parent plant – a large woody shrub actually – still performs royally every year.

S. guaranitica comes from South America and as this is near the Equator, it has very little variation in the timing of day and night throughout the year. So unlike all northern hemisphere herbaceous plants, the diminishing daylight does not limit its flowering, but cold will stop it in its tracks overnight.

The secret of keeping it going from year to year is to take cuttings in late summer from non-flowering stems – the shoots that break at 45 degrees root easiest – to plant out the following June. You can also successfully take cuttings from the new growth in April. Although it can happily survive a modest winter – especially if you have a southern, sheltered garden – I dig my parent plant up each November, cut it back hard and put it in a container in the greenhouse to rest, protected from frost. It does well on our clay loam and will grow to 3m (10ft) tall in a warm, wet year. We also have *Salvia* 'Purple Enigma' in the Jewel Garden, which is not quite so vigorous and has, as its name suggests, richly purple flowers.

If you want a true-blue salvia then *S. patens* is your best one. It is hardy, especially on well-drained soil, although a mulch of bracken or even just heaped up leaves will ensure its survival over winter. The stems are sticky and it has pale green leaves and the royal-blue flowers are carried in pairs. *Salvia* 'Guanajuato' has bigger flowers than the species – each bloom up to 8cm (3in) long, 'Cambridge Blue' has, as the name suggests, paler blue flowers, and 'White Trophy' is a white form, although this seems perverse to me as the whole point of *Salvia patens* is its glorious blueness.

Salvia patens is easily grown from seed, sown in spring under cover. It also forms tuberous roots, and these can be lifted and stored in a dark, frost-free place just like dahlia tubers. It is important not to let the tubers dry out and they are best stored in leaf mould or vermiculite, watered lightly and re-dampened every few weeks. They can then be potted up and grown on in a greenhouse from mid-February, dividing the tubers as you go to increase the plants.

I have *Salvia elegans* 'Scarlet Pineapple' growing in a large pot although for years I used to plant it out in the garden each spring and pot it back up again as the frosts appeared in autumn. This salvia really is a late-flowerer and a hit of frost reduces it to soggy black rags so a conservatory or porch is the best place for it. There, if you water and feed it over winter, it will keep on flowering until spring. We grow it in pots and move it back to the greenhouse as soon as the weather looks chilly where it flowers happily until Christmas. The leaves smell strongly and deliciously of pineapple and the flowers are a brilliant scarlet so it is worth the trouble of accommodating its unseasonal habits. It fares well in thin soil and needs much less water than *S. patens* or *S. guaranitica*. It is also exceptionally easy to root cuttings from the side shoots which, once rooted, are best left more or less unwatered in a coldframe or greenhouse until late spring and then potted on to enjoy whatever warmth they can get.

GOLD

As autumn takes hold and the last rays of summer sun become increasingly fleeting, the garden spends a few days drenched in gold. These are some of the most beautiful moments of the year and depend upon the low angle of the sun combined with a clarity of light and body of plants that you cannot get again. But you can plant and plan to bring as much gold into the garden to provide the same warmth and splendour throughout the gardening year.

Of course, substituting any old yellow when you really want gold is not going to do the trick. Yellow does not achieve the same effect at all. Chrome yellow is the brightest colour in the spectrum and seems to throw out light of its own. Gold has undertones of brown and orange and shimmers, reflecting light.

Salvia guaranitica *is one of those exotic, tropical plants that is robustly reliable at Longmeadow, thriving in our rich, damp soil – although it does need some warmth to flower.*

Generally, spring yellows tend to be paler than the ones at this time of year, so golden flowers are thin on the ground until midsummer.

However as summer progresses the choice increases. Marigolds (*Calendula*), the Californian poppy (*Eschscholzia*), *Lysimachia punctata*, *Achillea filipendulina* 'Gold Plate', the imperial fritillary, are all capable of adding gold to the garden. Heleniums are in general more marmalade colour than gold but *Helenium* 'Goldrausch' bucks that trend. Solidago is a coarse plant for all except the back of the border but I like it, especially in a shady corner where it can shine out.

Golden foliage is more constant throughout the growing season and therefore vital. The golden hop, *Humulus lupulus* 'Aureus', is more lime green than gold at first but sunshine and age help it to become as good as gold later in the year. I grow it up bean sticks in the border and it can be cut back to the ground when the leaves begin to brown.

A couple of hollies add a spiky golden touch to the winter border. There is *Ilex aquifolium* 'Golden Queen' which has a golden margin. Despite its name, it is male, and therefore produces no berries, whereas, *I. a.* 'J.C. van Tol' has green leaves but produces yellow berries. I think they can pass as gold. There is certainly no doubt about the fruits of the crab apple *Malus* 'Golden Hornet', which are beautiful little golden beads clinging to the branches.

Grasses, especially in this early autumn period, can make a garden glisten with gold and catch the light like no other group of plants. The wonderful waving wands of the giant stipas are molten gold as the evening sun hits the seedheads. *Calamagrostis* × *acutiflora* turns from an orangey brown to gold with the light behind it, and *Carex elata* 'Aurea' needs sun only to enhance rather than create its golden tones. *Miscanthus nepalensis* produces gold plumes like delicate strands of gold thread. These make wonderful cutflowers, opening out in the warmth of a room from zigzag streamers to fluffy feather dusters.

The oaten flowerheads of **Stipa gigantea** *catch and hold the evening light with a spectacular golden blaze.*

CYCLAMEN HEDERIFOLIUM

Cyclamen start flowering modestly in the shade of my Spring Garden in early August, their pinks and white out of kilter with the falling blaze of autumn and late summer. They are not the showiest of flowers but add a touch of lightness and delicacy that is unusual in the garden at this time of year.

They grow from corms that should be planted just below the surface in small groups, preferably in light soil, in the shade of shrubs or deciduous trees – although they will grow in the awkward dry shade beneath pines and other evergreens.

Cyclamen hederifolium has foliage marked with silvery veins that is lovely in its own right but does not compete with the flowers at all, appearing only after they have done their stuff. They will spread by self-sown seedlings that can be collected and replanted if new colonies are desired. It will tolerate any amount of summer drought and does not mind winter shade, so can be coupled with ferns such as hart's tongue, which shares its ability to thrive in dry, shady places. It needs some winter moisture to establish healthy leaves, which makes it an excellent companion for snowdrops, which also need some winter moisture.

COLCHICUMS

I suppose we are so used to flowers coyly peeping from a wrap of leaves that the quick emergence of colchicums unadorned by leaves makes them seem naked. In fact, their folk names are naked ladies or the politer meadow saffron. *Colchicum autumnale* is our only native colchicum and has big flowers followed in late winter by generous leaves that last into the next summer. It makes a good ground cover plant but not such a good neighbour if planted too close to delicate plants, so give it room to spread.

Like most bulbs, colchicums need a period of dry when dormant but do not really like too dry a site, preferring rich soil that mimics their natural meadow habitat. Give them full sun as the flowers will close up if shaded or overcast. As their vernacular name suggests, they grow well in grassland or in a lawn around a tree, but you must be careful not to cut off the emerging flower buds when you mow.

The seeds of orach, the giant purple spinach that is such an important annual in the Jewel Garden, turn from a rich plum colour to a bleached tawny as summer goes into autumn.

If grown alone, unsupported by grass or other plants, they have a tendency to terminally flop after rain. If grown in a border, gravel makes a much more aesthetically pleasing background to them than bare earth, even if it is only a thin mulch over the surface.

Colchicum speciosum is vigorous and will, in time, establish large drifts in grass or under shrubs. *C. speciosum* 'Album' is big and pure white and lasts particularly well, not succumbing to wet weather as readily as other colchicums. I have planted some in my Writing Garden and hopefully they will spread and in due course make a lovely snowy September drift there beneath the apple trees.

The time to plant bulbs of colchicum and autumn-flowering crocus is late summer or early autumn, so if you want them next year get on and plant some now, putting them in deep – at least 10cm (4in) – below the soil's surface.

BOX

We use box a great deal for low hedges and topiary – and used it a lot more before we removed eight hedges to accommodate wider paths in the Jewel Garden to create better access for a film crew. Over the years I have planted more and more of it, almost always generating new stock from cuttings, so that now I have nearly 1,000 metres of hedging and over 100 pieces of box topiary, all of which have to be tightly clipped (and the clippings gathered and tidied) at least once a year. Talk about making a rod for one's own back.

However tall or low you want it, no plant can be clipped and trained to such a convincingly solid evergreen hedge as the common box, *Buxus sempervirens,* whether a protective low barrier at knee height or a solid green wall towering over your head. It withstands an extraordinary range of climates and conditions. It grows slowly enough to need minimal attention and yet fast enough to be encouraging. It propagates from cuttings with laughable ease. I find that it adds vital substance and form to the garden, creating a framework for flowers in summer and adding a rich green structure in bleak winter months.

The common box grows so easily as a cutting that one healthy specimen can generate free plants by the score every year. This means that you can accumulate plants to use on a scale that invites a degree of playfulness or creativity. Large box topiary can be made from a number of plants growing together and mazes and parterres of great complexity have all been facilitated by the availability of the planting material.

Box has a mass of fibrous roots that are good at finding moisture and nutrients from wherever it finds itself and consequently it will grow in deep shade. However to get the healthiest, thickest plants it should be planted in good soil, making sure that the drainage is good, and in some sun. When setting out young plants to edge a border of any kind it is essential to keep them clear of surrounding growth to avoid them being swamped. In fact a tip learnt from hard experience is to grow cuttings on in a nursery bed until they are bushy and at least 30cm (12in) tall before putting them into their final positions edging a border. On the other hand, if they are to remain unchallenged for light, you can stick the actual cuttings where they are to grow, 15–23cm (6–9in) apart, as a kind of micro hedge. To make a dense hedge rooted plants of common box should be about 23–30cm (9–12in) apart and those of *Buxus* 'Suffruticosa' twice as close.

There are exceptions to this rule. Around 20 years ago, I bought my first section of box hedging for my garden through an advert in the local paper. It said: 'Box plants for sale. Bring own spade'. When I turned up, spade at the ready, the woman selling pointed to a long, very established hedge. So I dug it up and drove home with the boot – and the back and front seats – crammed with very earthy box plants. I could see that it was a large-leafed variety called 'Handsworthiensis' that grows rather upright and very vigorously. She had planted it very close together but I spaced each plant much wider – about 90cm (3ft) – transforming her thickset hedge into an avenue of slightly startled plants that stretched the length of the newly laid-out vegetable garden. However, I knew that within a few years they would spread and grow together and it is now so dense and substantial that we use it as a tabletop for clippers, mugs of tea and jerseys shed as the sun – or the work – heats up.

The winter of 2010 devastated my box hedges, particularly in the Jewel Garden. But this was made much worse by the fact that they had all had quite a hard trim just a few weeks before the really severe cold weather set in at the end of November. I had been lulled into a false sense of security by the previous very mild winters and had got into the habit of making the second annual trim later and later in the year. I shall not do it again! As a result, great sections of hedge – particularly in the centre of the plants – died back and had to be radically pruned. However, without exception, they are all growing back strongly and I would expect them to be totally recovered within 18 months. It was a lesson that even box is not totally hardy.

Psylla buxi is an aphid that lays its eggs in the box in late summer. In spring these hatch out and the grubs feed off the sap of the leaves before flying away in June. They are evident by the white, sticky dust that they exude and which puffs out of the box when you brush against it. They stunt the plant's growth causing the foliage to pucker and look like mini cabbage leaves. However, young plants seem less susceptible than mature ones and the aphids seem to cause no long-term damage. They seem to particularly like the 64 box balls in the eponymous Box Ball Yard. I suspect that this is because the soil there is very poor and the plants are therefore stressed and more susceptible to predation.

There are two serious fungal diseases of box, both collectively known as box blight. Both cause patches to go brown and defoliate and even die. *Volutella buxi* has been around for some while and causes pink pustules to develop beneath the leaves, which then turn yellow and fall off. *Cylindrocladium buxicola* was only recognised in the 1990s and is more serious, although there is evidence that plants can and do recover from an attack if action is taken. The indications are stunted leaves that develop dark spots and the interior of the plant is bare and has wispy areas of brown and grey fungus.

Both types of fungi need humid, damp conditions to thrive. If you see signs of either, remove all the affected parts and gather every scrap of leaf before burning them. Feed the recovering plant with a weak solution of liquid seaweed weekly for a year and do not trim it until it is growing healthily – at which point stop feeding, as lush growth is more susceptible to the blight than hardy, slower shoots.

Before cutting the box balls I always take a batch of cuttings and over the years this has generated thousands of free plants for box hedging.

ASTERS

Michaelmas falls on 29 September and traditionally this was a day of reckoning in the countryside. It was a quarter day when rents had to be paid, when harvest could be assessed and when the Feast of St Michael would be celebrated with a goose fattened on that year's corn. Even more significantly, it marks the definitive end of summer. Autumn is here and winter is coming.

So Michaelmas daisies – asters – are significant and evocative in this very particular season. They are not showy plants but are often cheerfully beautiful, sometimes sublime and always wonderful for attracting insects into the garden.

The easiest asters to grow are the novi-belgii cultivars. These are not Belgian at all – they come from New York – and for years had a bad name because of their proneness to mildew, but they do have a wonderful range of colours from the lilac 'Ada Ballard' and the double blue 'Marie Ballard' to the deep plum colour of 'Winston S. Churchill'.

There is no doubt that powdery mildew can be a problem with Michaelmas daisies – however lovely the flowers. The fungus starts on the underside of the leaves and then spreads rapidly, blown by the wind. Keeping all weeds down helps a lot as they act as hosts to the mildew. The fungi live through the winter months attached to the dried-up leaves and will almost certainly infect the young shoots in spring. Removing all top growth at the end of the flowering season and shredding and composting it will reduce spores surviving the winter. It also helps if you keep the plants properly watered and mulched as the mildew spreads quicker in drought. Lift the plants every three years and divide them before replanting the exterior sections in a fresh site (this can be just a few feet away from the original spot). By discarding the interior you will provoke healthy growth that will inherently be more mildew-resistant.

However, the New England asters, *Aster novae-angliae*, are pretty much pest- and disease-resistant. The downside is that there is not a wide choice of cultivars. 'Barr's Violet' is a good purple, as is 'Violetta'. 'Lou Williams' and 'Septemberrubin' are just on the pink side of plum. All are happiest in moist soil, so dig in plenty of organic matter when you plant them and mulch them extra thickly in spring. Like most asters they flower best in full sun.

All asters can easily be propagated by division. A dry day in early spring is ideal. Lift a clump and split them with a spade into segments about 15cm (6in) across. Plant these new divisions in small groups directly into soil that has been enriched with garden compost, with each piece about 60cm (2ft) from its neighbours. Water well and mulch thickly. You can also take cuttings in spring from fresh growth, making the cut right at the base of the shoot.

It is an especially good idea to regularly divide New England asters, not least because large clumps invariably become very dense and woody and intrusive into anything but the very largest border. Dividing them often also reduces this group's tendency to lose its lower leaves. However, it is a good idea to place all New England asters towards the middle or back of a border so that their naked lower regions can be decently clothed by other plants in the foreground.

I planted some *Aster umbellatus*, the flat-topped aster, in my garden this year and they have tall stems with a mass of small white flowers that go very well with grasses. I also put in some *Aster divaricatus*, one of the few asters that enjoys shade. Again, it has small white flowers but carried on much shorter stems and is a modest plant if grown alone but good if allowed to form a low-growing drift. It remains light and elegant despite effectively being ground cover.

Aster × frikartii likes the same alkaline, well-drained conditions as *A. amellus* (it is a hybrid between *A. amellus* and *A. thomsonii*), so use mushroom compost to mulch it with and avoid cow manure or composted bracken. It is otherwise problem-free. We grow 'Mönch', which is pale lavender with flowers that continue from July to October.

PLANTING SPRING BULBS

It is good practice to plant spring-flowering bulbs in September because after this the bulbs lose precious growing time and their subsequent performance is impaired. But it is more than that. It is an essential gesture of hope, throwing a lifeline across the gulf of winter to the excitement and promise of spring.

'Bulb' is a generic word to cover those plants that store their next season's flower, and the nourishment to grow it, within a self-contained capsule that can survive until the following growing season without nourishment or roots. Everything about the flower, everything it ever needs to know in order to become its full self is contained within the squat bulb.

True bulbs – such as tulips, alliums and daffodils – are essentially a much-reduced stem made from concentric layers of fleshy scales with a protective dry outer layer. Each scale is either the base of a leaf or the thick scale leaves that never appear above ground. The papery outer layer is what is left of last year's scales. Most bulbs are smooth, but some lilies and fritillaries have no protective skin and the scales are separate.

There are three distinct forms of bulb. In the most common – like tulips or alliums – the bulb shrivels and dies after flowering and is renewed from buds formed at the base of the scales at the point where they join the basal plate.

Narcissus bulbs, on the other hand, do continue year on year, producing offsets rather than wholly renewing. This is why you get ever-increasing drifts of daffodils whereas tulips tend to increase more reluctantly and with a marked loss of vigour – as it takes two or three years for most tulip bulbs to flower.

The final type of bulb is like a hippeastrum (*Amaryllis*) and has embryonic bulbs for three years ahead within each 'parent' bulb – so it is genuinely perennial.

One of the advantages of planting spring bulbs now is that you are forced to work around plants so you are less likely to find yourself digging the things up later in the year when you are planting or cleaning up the border. There are two rules when planting bulbs. The first is to allow at least twice their own depth of soil above them and the second to put them pointy end up. But as with all good rules, there are exceptions – the imperial fritillary for example needs to be fully 15–23cm (6–9in) deep and planted on its side, and although daffodils especially will respond to shallow planting by not flowering, tulips can be planted just below the surface if they are to be lifted later. However, the guiding rule is that you will do less harm by planting too deep than too shallow.

Another general rule is that bulbs need good drainage and this is vital for tulips, alliums and *Iris reticulata*. The best way to ensure this is to mix in loads of grit into a general area or container (50:50 grit to potting compost is none too gritty) that they are to be planted in, or to add a good dollop of grit in every planting hole. Tedious, but worth it. As ever, there are exceptions to this rule, too. Snowdrops and the snake's head fritillary not only prefer damp conditions but also need them to thrive. It is an oft-repeated piece of advice but snowdrops are best spread 'in the green' which means digging a clump up and dividing it thinly either whilst they are still in flower or just after. They can, of course, be planted as bulbs, but they are very small and fiddly and you will have a much better success rate if you establish them as plants.

Planting narcissi in the Orchard. The problem with planting bulbs in September is that the ground is usually at its driest and therefore rock hard.

When planting bulbs into grass you can either take out the holes individually for larger bulbs such as daffodils or, for little bulbs like crocus, cut squares of turf, place the bulbs on the bare soil and put the turf back over the top of them. I have planted thousands of crocus like this and it works very well. The best and easiest way to plant larger bulbs in grass so that they make a natural spread is to throw a handful onto the ground and plant them wherever they land. Of course, you must never mow the grass until the leaves of the bulbs have started to yellow and die down, which, for daffodils, is going to be as late as mid-June.

As a rule it is a good idea to plant all bulbs as soon as you buy them because the risk of fungal disease and rotting increases the longer they are stored. Some are much worse than others – I once bought 200 erythronium bulbs and only 50 survived the two weeks that I delayed unpacking them.

All bulbs look good in a container, particularly terracotta pots. These can be moved into sunshine or shade as required and can also be guaranteed to provide the right kind of drainage. Pack smaller bulbs in tightly and try to find shallow alpine pans that are perfect for crocus, snowdrops, fritillaries, iris, muscari and other small bulbs. Remember that most will need a summer baking so make sure they have some sunshine when dormant.

Bulb Table

PLANT	TYPE	SOIL PREFERENCE	PLANTING TIM	FLOWERING
Aconite	Tuber	Rich, shaded	March	Late winter
Allium	Bulb	Well-drained, full sun	Autumn	Summer
Anemone blanda	Tuber	Well-drained	Autumn	Spring
Anemone nemorosa	Rhizome	Well-drained	Autumn	Spring
Arum	Tuber	Moist, well-drained	Autumn	Spring
Crocus	Corm	Well-drained	Autumn	Spring
Daffodil	Bulb	Well-drained	Autumn	Spring
Erythronium	Tuber	Cool, well-drained	Autumn	Spring
Freesia	Corm	Well-drained	Autumn	Early spring
Fritillaria meleagris	Bulb	Damp	Autumn	Spring
Hyacinth	Bulb	Well-drained	Autumn	Spring
Iris	Rhizome	Varied	Spring	Spring to winter
Lily	Bulb	Well-drained	Autumn	Summer
Muscari	Bulb	Well-drained	Autumn	Spring
Nerine	Bulb	Sandy soil	Autumn	Spring
Scilla	Bulb	Well-drained	Autumn	Spring
Snowdrop	Bulb	Moist	After flowering	Winter
Tulip	Bulb	Well-drained	November	Spring

AUBERGINES

I could happily eat ratatouille every day for the rest of my life, with onions, garlic, courgettes (or squash), tomatoes and herbs all grown and harvested in the garden – and of course aubergine. I grow aubergines and have done so for the past 20 years and yet I always do so with some qualms.

The key to growing any crop is the harvest that it produces and all but one of the above crops usually harvest generously and well. But I confess that most years I struggle to get a decent crop of aubergines. If they are good then we do not have enough. It becomes a case of celebrating that we have grown them at all rather than that they are seasonal, fresh and truly delicious – which are the standards by which all food should be judged, whether grown in the garden or bought in the fanciest restaurant.

The last time that they performed well for me was in 2006 when we had the last truly hot summer. I grew them with tomatoes in the top greenhouse – after all they are cousins, both from the Solanaceae family – and they effortlessly provided scores of lovely glossy purple fruits. And because it was so hot, I watered every day. In a year like 2011 that would create a soft, warm humidity that would be the perfect recipe for tomato leaf mould – and any other fungus going – but in the fierce heat of 2006 it was just enough to keep these thirsty plants happy. Moreover, because it was so hot I opened every door, window and vent and often left them open all night – which did little to help the aubergines but stopped the tomatoes getting too humid.

I now grow aubergines in pots standing on the heated propagating bench and subject them to the same regime as the cucumbers and melons. They like it and thrive but would produce more fruit if they were grown a little harder – ie had more air heat.

I sow the seed in March, pricking out the surprisingly modest seedlings into 8cm (3in) pots and then in early June when they have a decent root system, into the 30cm (12in) pots they remain in. If I had more space I would give them double the pot size. I give them the same potting mix as cucumbers – with lots of extra garden compost but plenty of grit to create a loose root run.

They grow about 90cm (3ft) tall and I give each a bamboo stake to stop them being bowed down by the weight of the fruits – although that is based as much on hope as on experience. The key is to pinch out new side shoots after four or five fruits have formed otherwise you get a bushy, luxurious plant and no extra fruit.

Slugs and snails love the fruits and will hide up in the leaves to gouge into the flesh. The fruits can be slow to develop but should not be left too long. As soon as they are a reasonable size and shine like a guardsman's boots then they are ready. Overripe fruits turn dull and taste impossibly bitter.

CHICORY

I like the way that chicory has two distinct phases of growth and with some of them the changes from one to the other can be dramatic. Part one of their cycle, between May and September, is to establish a deep taproot, and they will sprout lots of green leaves towards that end. But these leaves are impossibly bitter and tough – and in quite a few varieties surprisingly furry. Not appetising. However, that is not their role. They serve to feed the taproot and supply the gardener with composting material. The only thing to watch, as they put on this summer greenery, is that they do not become too crowded as unlike lettuce, which always does well in wet conditions, chicory hates wet air and very quickly rots so they need generous spacing and also regular removal of excess foliage.

Around mid-August I start to strip away the outer leaves to let air and light in between them. By the beginning of October this process is finished – barrowloads going to the compost heap – and what is left is stage two, the harvestable plant which responds to temperature and light levels to change leaf shape and sometimes colour. However, in most chicories this is just the first of two or even three such harvests as a decent taproot will produce plenty more leaves after cutting.

For years the British – if they were feeling adventurous – grew just 'witloof' chicory, which needed blanching under pots or in a cellar. But radicchio in all its assorted glory is adored by the Italians who grow it with great care and expertise, especially in the Veneto. I remember just ten years ago having to go as far as Venice to find the seeds to

(Following pages, left) 'Rossa di Treviso' has upright deep crimson winter foliage that curls in on itself like the tail feathers of an exotic bird.

(Following pages, right) 'Rossa di Verona' has looser, more cabbage-like leaves, like a butterhead lettuce, but also develops gorgeous ruby-coloured leaves in autumn.

grow a decent range of types. Since then I have grown 'Rossa di Trento', 'Rossa di Verona', 'Grumolo Verde', 'Catalognia', 'Variegatta di Chioggia', 'Castelfranco', 'Minutina o erbe stella' and others. They have mostly been experiments and over the decade most have fallen by the wayside. But I do still grow 'Rossa di Verona' which is slightly speckled and makes an almost cabbage-sized head, and 'Palla Rossa'. I also love 'Rossa di Treviso' which is upright and cos lettuce shaped with wonderful curling alizarin leaves that have clear white ribs. This is perfectly good raw – a bitter, grown up taste like 80 per cent dark chocolate or very dry fino sherry – but even better cooked.

The red chicories turn from green to red in response to daylight length and temperature. The winter leaves can be cut flush with the soil and will regrow to provide three or four crops before spring. The plants do not mind cold but if the weather is very wet an open-ended cloche will stop the leaves rotting.

I also always grow endive and nearly always just the curly type like frisee rather than escarole that has almost flat leaves. Occasionally I tie them up to blanch them but then they fill with slugs. Better to plant them close so they blanch themselves. Unblanched they are spectacularly bitter, although if they survive frost they are the sweeter for it. Unlike chicory (which needs a long growing season from April to March) endive will be ready for harvesting in a few months, so can be sown from January to August for a year-round supply.

CITRUS

I have had a couple of oranges growing in pots for the past 20 years. In late September I take these back into the shelter of the greenhouse in case we have an early frost. The key to their winter health is plenty of air, some light and not too much moisture around their roots but moist air.

Remember these things when storing citrus plants inside from September until May. Most houses are far too dry and hot for overwintering citrus. The leaves will start yellowing, curling up and ultimately dropping off. I have found that a cool greenhouse kept from 5–10°C (41–50°F) is ideal. The important thing is to protect them from big or sudden changes in temperature. Interestingly, oranges grown in cool climates develop much brighter colour than their counterparts grown in the sun – although the latter are juicier and sweeter.

The secret of potting compost for any citrus plant is drainage. Yellowing leaves or, worst of all, phytophthora (manifested by a gummy exudation from the stem) are usually a result of too much water. They hate sitting in waterlogged soil so I add plenty of horticultural grit to my homemade potting compost. I topdress this with a thick mulch of good garden compost, although well-rotted animal manure would probably do as well.

The ideal time to repot is in May or June and should be done every two or three years, putting them into slightly bigger pots each time or, as is the case with my two 20-year-old citruses, trim the roots back slightly and repot them with fresh compost in their old container.

It is important to leave a generous rim above the layer of mulch so that when you water you can really soak the plant until it is pouring out of the drainage holes at the bottom. If the compost is adequately free-draining almost all the water should drain very fast. In winter there is no need to do this more than once a month and in summer no more than once a week. It is much better to use rainwater if available but tapwater is fine if left standing for 24 hours for the chlorine to evaporate. In any event, remember that a good soak every now and then is much better than frequent light watering. It is also a good idea to stand them on blocks of some kind to avoid any risk of the pot sitting in a puddle after watering.

Seville oranges, *Citrus aurantium*, are very hardy and have a wonderful fragrance although the fruit are not edible raw. Sweet oranges, *Citrus sinensis*, can be grown – and eaten – but can be a little touchy about cold. But they are not nearly as sensitive to cold as limes, *Citrus aurantiifolia*, which start to get very unhappy when the temperature drops below 5°C (41°F).

Perhaps the easiest citrus to grow in a pot are lemons, *Citrus limon*. These are typically vigorous and spiny and have the great virtue of holding three crops at once. At any one time there will be flowers, young green fruits and ripening yellow ones on the same tree. This means that with some careful management you can have a constant supply of lemons from one tree. New shoots of lemon trees are typically purple in colour, as indeed are the buds.

Citrus plants can take huge amounts of pruning and it is best to do this in spring when they go outdoors, thinning the centre of the plant so that light and air can get in. Take away the branches that have produced any fruit over winter and any that are looking tired and lacking vigour. The harder you cut, the greater the health and vigour of any remaining branches, so do not be timid.

Although the fruits are decorative and can remain on a mature plant for months, it is advisable to remove most from young plants as they take up a lot of energy. Removing ripe fruit also stimulates flower production, followed by even more fruit.

Scale insects (*Planococcus citri*) thrive in warm, humid conditions and are often seen in late winter or early spring. The first sign is likely to be the sooty, sticky residue that they leave on the surface of the leaf below them. Turn over the leaves above and you will see the scale insects sticking like limpets and extracting the sap. The best way to deal with this is to wash the leaves every now and then with a soapy solution. A foliar feed of liquid seaweed in early spring will do wonders to restore the health of an afflicted plant.

FIGS

The figs in the Walled Garden suffered horribly in the extreme cold at both ends of 2010. Not only did the air temperature get down to -15°C (5°F) here but there was a wind with it. This is hostile territory for figs and whereas they looked battered and shocked in the summer of 2010, as spring unfolded in 2011 it was apparent that they had suffered grievous and permanent damage. It looked like we had lost them all.

This was extremely sad as I planted all 10 on my fourtieth birthday, on 8 July 1995. All had grown lustily and borne us good fruit and had become an important architectural element of the Walled Garden. So their loss was a big blow.

However, against all advice, I did not cut them down but waited to see if there was any regrowth. Throughout spring they showed no sign of life but in July they started sprouting. By August it was clear that they were all alive and throwing up strong shoots both from the base and, almost at random, from the dead-looking trunks and branches. Eventually, in mid-August, I cut them back hard – but this was a case of pruning rather than felling.

I love everything about fig trees, from elephant trunk stems with their curious wrinkled bark that looks as though it has rucked and slipped a little down the wood it encases, to the huge modesty-concealing leaves and, of course, the voluptuary promise of the fruit. This is their season, producing deliciously succulent fruits through into next month. A fresh fig, still warm from the sun, ripe enough to split apart with your thumbs, is a life-enhancing thing.

Citrus plants, whether orange or lemon, look best in terracotta pots. This also means that they can be brought under cover and protected from winter frosts.

Figs can produce three crops simultaneously and invariably have two on the go at any moment. At this time of year there will be the large ripening figs, half-sized ones and, if you look closely, tiny pea-sized, even pin-head, fruit tucked into a joint between stem and leaf. These tiny ones are next year's harvest. The in-between ones – essentially any that do not ripen by the middle of October – will never ripen in northern Europe although further south they will produce a delicious harvest from New Year to early spring.

In a mild winter in this country they start into further growth but the skins are tough, so they split and fall off, which weakens the plant and delays the growth of the very small ones, meaning that they often fail to ripen, too. The solution is to wait until you have harvested the last ripe fig at the end of October and then remove every single fig bigger than a fingernail. There is a huge temptation to leave some, but it should be resisted. Put your trust in the minute figlets that will grow and swell very fast the following summer.

The fruit are formed towards the tips of healthy young shoots so for a maximum harvest it is best to roughly fan-train the fig against a wall, removing about a quarter of the oldest stems every year along with any growth that is growing out from the wall or crossing. Do this in April. Then in August, prune away any overly vigorous outward growth that will shade the ripening fruit.

There is a myth tagged onto figs that if they are to fruit at all, they must be planted into a rubble-filled hole. This is not true. Figs need a decent start in life if they are to become a healthy tree. But some restriction on the roots will limit the growth of the tree and stimulate a higher rate of fruit. The answer is to plant a fig against a south-facing wall if you have such a thing to hand. The southern aspect will give it maximum sunshine and the wall will significantly retain the heat around the plant to improve and prolong ripening. If you do not have a wall to plant against the answer is to create an open-bottomed box to plant it into. Old paving slabs are ideal for this. Leave enough room for the fig to get established in good soil before the roots butt up against the slabs. The slabs do not have to be closed tight together – some constriction is all it needs.

Ficus carica 'Brown Turkey' is the variety that you are most likely to find in a garden centre and it is certainly most likely to ripen outdoors in all but the warmest and most protected British gardens. It is a good tree and delicious fig but by no means the only one. 'White Ischia' is, in fact, pale green and will grow well in a container under cover. 'Brunswick' has large green fruits that tinge brown. It can be grown outside, although needs a good hot summer to ripen. 'White Marseilles' is another potential outdoor fig that is early to ripen and the large fruits shade yellow when ripe. 'Rouge de Bordeaux' has small purple fruits with ruby flesh and it would be my first choice for growing against a brick wall in a greenhouse.

HAZEL NUTS

Hazel wood is the main source of bean and pea sticks, and all the fencing at Longmeadow is woven from coppiced hazel. But at this time of year the hazels exist for their nuts. Growing nuts for deliberate harvest is rare in modern gardens, but until the Second World War a 'nuttery' was a feature of many large gardens, and I have seen miles of hazel plantations in Italy, north of Rome, where the nuts are harvested for their oil. The biggest nuts are produced on old wood, so the hazels are pruned to get the right amount of light and air to them, fixing them in a perpetual maturity, whereas coppiced wood is always reinventing itself.

The wild hazel you find in woodland is *Corylus avellana*, or the cob. It carries its nuts in clusters of two to four, the squat nut sleeved by a short husk. To get the best nuts, hazel, rather like fig, needs to grow on poor, stony soil. If the ground is too rich and damp the goodness will go into the wood rather than the fruit. Hazel nuts will store until Christmas if kept in their husk, which stops them from drying out too fast. If collected whilst green, leave them in a dry place and they will ripen by about November.

PEARS

Once we reach September, the pears are ripening by the hour as the swallows frantically work the daylight sky above the garden, building up stores for their astonishing journey south. I know that every day is potentially the last for both of them and that general sense of harvest and subsequent loss tinges this season with melancholy.

I adore swallows. When the first one arrived in the garden last spring after such a long and miserably cold, wet winter the sight of a single swallow chirping and scything through the April air was enough to wipe away all winter blues. No long-lost friend has ever been more welcome. They often swoop into our house through any of

the doors left open into the garden from morning till night, check us out for insects and then just as elegantly swoop outside again.

The pears have been swelling for the past month or so and the earliest – 'Williams' Bon Chrétien' in the Ornamental Vegetable Garden – are all gathered in by the end of the month and half eaten.

As a general rule, the later an apple ripens the longer it will keep. Pick an apple at the moment when it will gently come away in your hand and it will be both perfectly ready to eat and will store in the best possible condition for eating months later. But pears must be picked before they are fully ripe and the ripening process is the very thing that limits their storage time. This is why it is almost impossible to buy a ripe pear from anything other than a good greengrocer or market stall.

A pear will ripen from the inside and gradually the flesh will soften and be at its juiciest best to eat until it reaches the skin. This is the moment when it must be eaten. But you have to take the opportunity exactly as it comes because the next stage is rapid disintegration and decomposition and the perfectly ripe fruit becomes a rotten, squidgy shell in a matter of days – if not hours.

Pears are a luxury – a seasonal treat. Better to eat a few of the real thing every year (that you have nurtured and been acquainted with from blossom to ripening) than a shiny, growth-regulated, tasteless affair every day.

Apart from the fruit, pear trees are glorious objects. A mature one can stand 15m (50ft) tall and in full early April blossom makes the best flowering plant on the planet. We have a perry pear in the Orchard that produces thousands of tiny rock-hard fruit every year but which is rapidly becoming a superb, flowering tree.

The ones that I grow are limited (so far...) to a handful. There is 'Conference' (1770) which is found growing against a sunny wall on almost every farmhouse in this part of the world and is the safest and most reliable of pears you can grow. In a bad year it is reliable because it will set fruit parthenocarpically – ie self-pollinate – although these fruit are never as good as cross-pollinated ones.

These 'Concorde' pears are windfalls gathered before the wasps can eat them. Although they will not store, most will ripen off the tree and taste delicious.

'Doyenné du Comice' (1894) is agreed to be the prince of pears, defining the essence of peariness. Because the fruit are so good, they are best grown as espaliers – which will maximise fruit production against tree size, and justify pride of place against a south-facing wall.

'Concorde' (1977) is a cross between the previous two and makes a beautifully healthy, archetypally pear-shaped fruit. It is very good but does not quite have the character of either of its parents. I grow it as a standard.

'Williams' Bon Chrétien' (1770) is used commercially for processing but is a good dessert pear. It is big and yellow with a honeyed, musky flavour. It is also one of the very first pears to ripen and all mine were ready by the first week of September in the exceptionally early harvest year of 2011.

Pear trees are more tolerant of cold and wet than apples and, especially if grown on their own roots, fantastically long-lived and hardy, although modern trees are invariably grown on a quince rootstock. This produces a smaller, earlier-fruiting tree, with Quince 'A' moderately vigorous and Quince 'C' more dwarfing.

The pruning regime for pears is very similar to apples, although once it starts to fruit a pear will take much harder cutting back than an apple.

The principles of pollination are the same for both apples and pears. Varieties are grouped according to their period of flowering. Obviously only two trees that are simultaneously in flower can perform successfully. So you need another tree from the same or an adjacent group to provide flowers that will overlap with your original tree's period of flowering. Whilst there are eight groups of apples, there are only four of pear, so the chances of accidental pollination from any pear in the vicinity are quite high. Of the best-known varieties, 'Conference' and 'Williams' Bon Chrétien' are both in group 3 and 'Beth' and 'Doyenné du Comice' are both in group 4. Obviously, any permutation of these will work.

Apple and pear canker is bad news. It is caused by the fungus *Nectria galligena*, which spreads in spring into any wound or opening. You are supposed to cut out all signs of it, going back into healthy wood and burning all prunings, but all the espaliers in the Ornamental Vegetable Garden have had it for years and my experience is that it invariably comes back, despite the tree regrowing vigorously. The problem is as much in cultivation as anything else. As with all fungal problems, good drainage and ventilation are essential. 'Concorde' seems to be more resistant, as does 'Conference'.

PRUNING RASPBERRIES

Pruning summer raspberries is a process that I really enjoy, especially if I can do it slowly whilst listening to a good play on the radio. It transforms a tangle of arching, overgrown canes into a perfectly ordered, ship-shape mesh, all ready for next year's action. It can be done any time after the last fruit but should certainly be completed by the end of September, not least because the longer you leave it the harder it is to distinguish between the old, brown canes and the fresh, green ones.

The process begins by cutting back to the ground all of the brown canes that produced fruit this year. This will leave you with a thicket of new, green canes that are ready for next year's crop.

I then reduce these so that they are evenly spaced and with no more than half a dozen per plant. Having done this I tie them to the permanent framework of wires spaced about 60cm (2ft) apart and stretched between really strong supporting posts every 2m (6ft). They could all be tied individually but I find it easier to weave a long length of tarred twine (which will last securely until next September) around and between each stem, fanning out them equally as I go, so that they are fixed tightly in position. This way they will not be moved and damaged by high winds, snow or – hopefully – a heavy crop of fruit next year.

Autumn raspberries bear fruit from late August into November and their pruning regime is much more straightforward: all top growth is simply cut to the ground around Christmas time, as they bear their fruit on the new canes produced each spring.

The apple 'Herefordshire Beefing' is an old cooking variety with a very low moisture content that was specifically grown for drying. However, it stores and cooks very well as it is.

WORCESTER PEARMAIN

The first apple to ripen in my garden is 'Worcester Pearmain' which has a lovely honeyed, almost fragrant flavour throughout September.

It first appeared in the 1870s in the 157-acre nursery of Richard Smith, at St John's, just on the outskirts of Worcester. This nursery claimed to be the largest in the world at the time with 200 staff and 18 miles of pathways. The nursery closed in 1993 but I remember it well.

The apple was an immediate commercial success, being awarded a first class certificate by the Royal Horticultural Society in 1875. The original seedling seems to have originated in the garden of William Hale in Worcester, close to Richard Smith's nursery. There were two new seedlings, one with bright yellow fruit, and the other with bright red. Richard Smith offered Hale £10 – around £1,000 today – for exclusive rights to take grafts from the tree. However Smith soon got his money back because within five years he was selling the trees like hot cakes at one guinea each.

Smith named the tree 'Worcester Pearmain' from the city and from the shape of the apple – 'Pearmain' apples have a slightly pear-shaped narrowing at the base.'Worcester Pearmain' is one of the parents for both 'Discovery' and 'Lord Lambourne'.

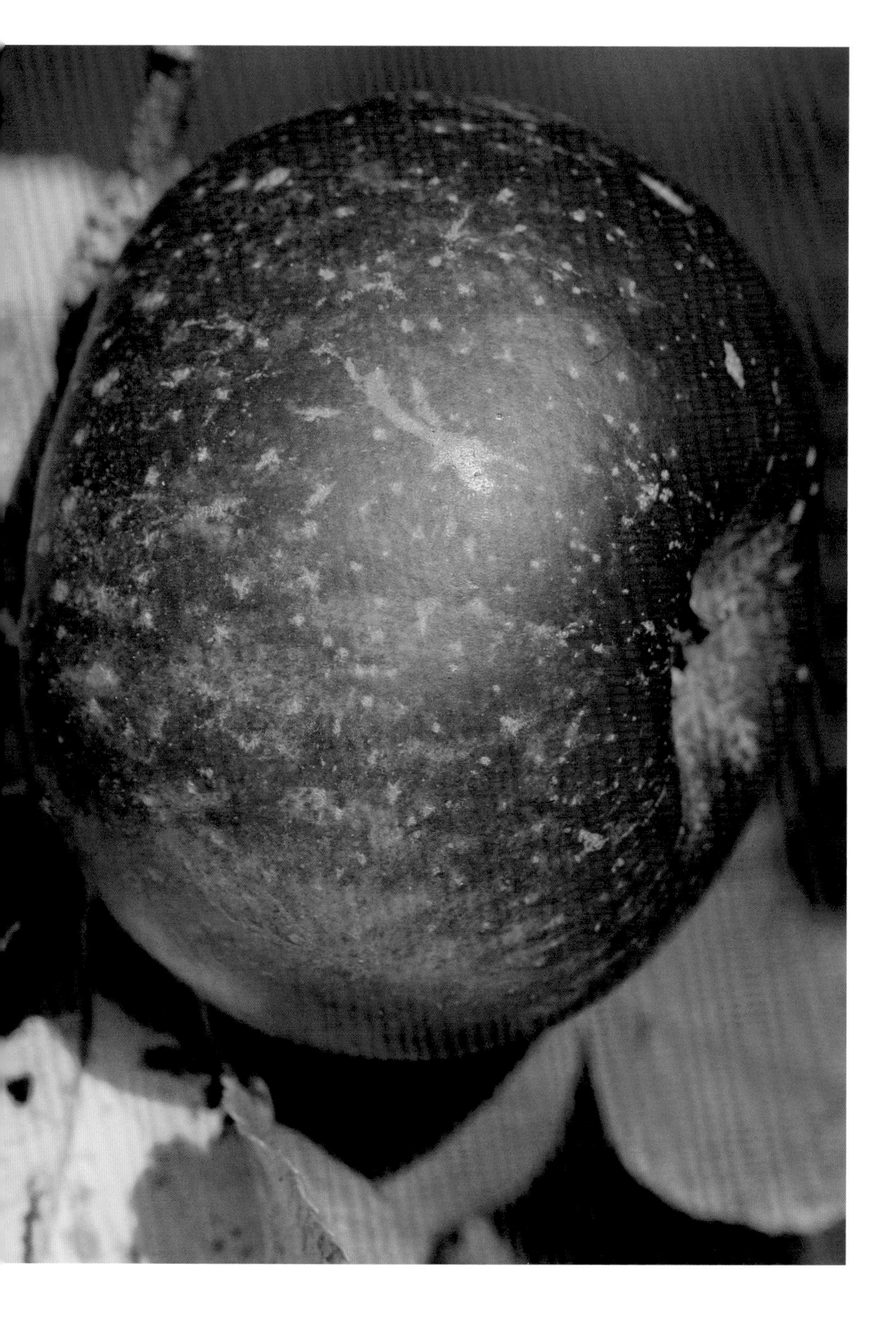

Foraging

I see my garden as a clearing made in the landscape – in time as well as space. In effect I am borrowing it from the countryside all around and in time it will surely be returned. 'My' garden will be lost but it will be filled with wildflowers and just as certainly full of 'wild' food.

The best-known and easiest harvest to forage at this time of year is the blackberry, *Rubus fruticosus*, and we have been plucking them from their prickly canes since Neolithic times. There was a time when you would not have left the garden to do this, for in the sixteenth and seventeenth century blackberries were more highly prized than raspberries.

There are wild raspberries, *Rubus idaeus*, in Britain but I confess that I have never eaten them. Just as blackberries were cultivated carefully, so the raspberry was considered a handy free fruit, more picked for its medicinal virtues in treating a sore throat than its taste. It will grow vigorously following woodland clearance or coppicing, ripening around midsummer, so September is at the end of its season.

There are wild plums in thousands of hedgerows along the Welsh border and in the West Midlands all growing on conveniently low trees that need no groveling – or indeed walking boots – to reach.

Bullaces, *Prunus domestica*, are thought to be native and are similar to (but rounder than) damsons. (Although, just to confuse things, some damsons are rounder than others.) Sloes are the fruit of the very native blackthorn, *Prunus spinosa*, and are mouth-shrivellingly astringent until ripe – which often is not until late October. But collected in September, pricked and soaked in gin and sugar they become sloe gin – the best Christmas liqueur.

In my childhood we would often get up before breakfast to go mushrooming – either in the meadows where cattle had been grazing or in the stubble fields – for field mushrooms that some years grew by the thousand and in others were as rare as four-leaf clovers. But it was always worth getting shoes soaked with dew and early starts because fresh mushrooms fried in bacon fat or butter for breakfast make any commercially grown version a poor imitation of the real thing.

The truth is that we British are not at all brave when it comes to mushrooms, whereas in mainland Europe they celebrate them and their season with the fervour that we show for strawberries or Christmas. The ones that are easiest, safest and always delicious are those found in the meadows around Longmeadow – field mushrooms, the delicately exquisite common ink caps and the giant puffball, which we eat sliced into thick fungi steaks and dipped in egg and breadcrumbs and fried. The best time to look for these is in autumn and 2011 was an astonishing year for these mushrooms. We gathered basketfuls every day for weeks, right into November.

Topiary

One job that I do at the end of August and beginning of September is to cut my yew hedges and topiary. Topiary is just another form of pruning and the golden rules of pruning apply. The first is to always use very sharp cutting implements. This is safer for you and better for the plant. The second is that the harder a stem is cut back, the more vigorous its responding growth will be. So pruning topiary creates a denser shape and more vigorous plant. Therefore if a shape has a hole or some weak growth it is best to go against your instincts and to cut this a bit harder than the rest to stimulate more growth.

Topiary trimming is a ritual that I look forward to. I have done it many times with shears, which are cheap, effective, energy-saving and good for the muscles of your forearms. But they must be very sharp. I also use electric cutters, both mains- and battery-powered, which are lightweight but powerful. In the past I have used a big petrol-driven hedge cutter but this is tiring, too heavy and noisy for something that needs concentration and a lot of reflection. It is too easy to get involved in what you are doing and forget to stand back and check that you have not cut too enthusiastically.

Although the yew cones are dramatic and dominate the whole of my front garden, they are surprisingly easy to train and trim. I start by cutting a straight line up the side of the cone, removing all surplus growth (and they always grow unevenly) and then keep moving round, following that line, working from bottom to top, until I go full circle. After that it is just a question of standing back, checking it, and then making good here and there. I can do all 26 in a few hours, rake up the trimmings and that is it for the whole year – other than occasional weeding around the roots and perhaps a mulch of compost in spring to give them a boost as they start to grow.

(Following page) The yew cones in front of the house have an annual clip every September which holds them trim right through to the following summer.

Unlike box, yew is slow to get growing in spring but can easily put on 23–30cm (9–12in) by midsummer, so it doesn't take long to develop a really good specimen from an initially lax bush.

There is no mystery to cutting the box balls. This is mainly because they are not balls at all but rounded cobbles. Each one is different and has its own shape and character. The effect is made by their accumulated and repeated shapes rather than by any single one of them, so there is no need to be precious about it. I simply start at the top and cut in a curve downwards and across, moving steadily round.

There is a tendency for the rounded curves to become domes, tethered to the ground on a wider base than is necessary or aesthetically desirable, so they need undercutting and flattening on the top. I find it easier to keep cutting until it seems to be right, rather than pecking away at it, and the less I think about how to do it the better it gets done. Actually this bit of the garden, with its apparent rigid grid and formality of 64 topiary box cobbles, is all about feeling right rather than exact measurement.

Box cuttings

Box propagates very readily from cuttings and September is the perfect time to obtain hundreds of free box plants. Choose healthy, vigorous shoots and cut just below the point where this year's growth begins, so that there is a plug of older wood at the base. This stops them drying out too fast. Strip off the bottom half of the leaves and trim back any bushy side shoots. You can either stick them into the ground in rows, digging in plenty of sharp sand or horticultural grit to encourage drainage and good root growth, or into containers with a potting compost mixed 50:50 with grit or perlite.

If placed directly in the ground (a corner of the vegetable garden is ideal) they can be left for up to two years before transplanting. If in a container, they should be lightly watered and then left in a sheltered position outside or in a coldframe until the following May, and ideally given a full year to develop a healthy root system before transplanting to individual pots or a border.

BOX CUTTINGS

Choose healthy new shoots, cutting just below the new growth so there is a base of harder wood.

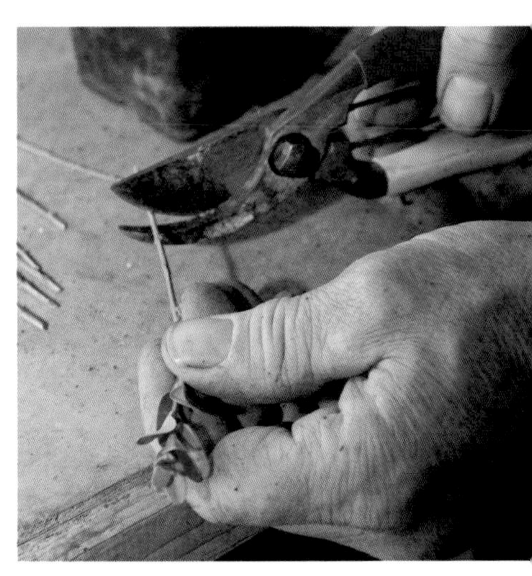

Trim each cutting just below a leaf node, using sharp secateurs or a knife, so they are all the same length and remove at least half of the leaves.

Fill a plastic pot with a gritty cutting compost mix and gently insert the cuttings around the outside between the pot and compost so the bare stem is fully buried.

Label them clearly with the date, water them and place the pots on a heated bench (if you have one) or in the corner of a greenhouse or coldframe.

Autumn lawn care

In September I start to prepare my lawn for spring. The first thing I do is to give it a really good scratch. The best implement for this is a wire rake with spring tines. This will pull out all the 'thatch' or the dead grass and roots that inevitably accumulate during the growing season. It will also loosen any moss you might have. Moss is caused by poor drainage and shade. Deal with these underlying problems and you have a long-term solution to moss in your lawn.

I gather all this up – and you will be surprised by how much there is to gather – and add it to the compost heap. This can leave a very bald-looking lawn, but it quickly recovers. Any really bare patches can be sprinkled with grass seed, which will germinate this month and be thick and ready to cut next spring.

Then I give it a good spiking. This is tedious and hard work, especially if the ground is hard. On a small lawn or a path, a fork is ideal but on a larger area it might be worth hiring a mechanical machine or even buying one of the many gadgets that will do it. The result is worth the trouble because it opens up the compacted soil, and all lawns are compacted by the end of summer simply through feet walking over them. Aerating the lawn lets air in, loosens the ground for the roots to grow and allows the rain to drain.

The next phase is to brush sharp sand into the holes made to improve drainage. This might seem counter-intuitive but will help enormously becauseit will encourage the roots of the grass to grow deeper and therefore be better equipped to search for moisture if a drought occurs. Finally, I raise the blades on the mower for the rest of the autumn. It is much better for the grass to go into winter with a protective coat.

OCTOBER

The beginning of October is my gardening new year. Summer is over. The harvest is in. Time to take stock, plan, prepare and start again.

But I am always surprised at how slowly autumn creeps in. The drift from September often flatters to deceive. The flower borders still have their glowing richness that the first frost or wet storm – and one of these two is almost guaranteed this month, if not both in the same week – will brutally snatch away and are filled with russets and crimsons, purples and oranges, and the refreshed grass often has an emerald intensity that is lost in the dry days of summer. It is an earthy time with the smells of fruit, worm casts and falling leaves sifting through the air like wood smoke.

Yet, by the end of the month, autumn is always unequivocally there. The leaves are changing colour daily and the air has an almost tangible opacity you only find in autumn. It cannot be withheld or denied. The sound of winter's marching footsteps can be heard over the horizon and finally, on the last weekend of the month, the clocks change and a door closes tight until next spring.

Because this is borrowed time, because the frosts will come any day, these days are as precious as jewels. It is too dark and cold to eat out in the evening but there are still meals in the garden in the middle of the day and October sun and light before the first frosts are as beautiful as any other time in the garden. I like the way that the season can shift almost overnight in October with soft days glowing with light and midday sun warm enough for shirtsleeves and then blasts of weather that send you huddling for warmth and shelter. However, it is a very busy month with a lot to do before the bad weather sets in properly. The Ornamental Vegetable Garden is still very productive at the start of the month with the shortening light limiting growth as much as the temperature, and in the first half of the month we have lettuces, rocket, mizuna, runner beans, squashes, fennel, spinach, chard, beetroot, sweetcorn, carrots, turnips, the first parsnips and maincrop potatoes.

We mow the grass until it is too wet to do so and make sure that all the hedges are trimmed so that they go into winter with their crisp outlines revealed, holding the garden together on even the drabbest days. But we never take anything out of the borders or cut anything back in October unless it is ailing badly because the birds love the seedheads and the slightly unruly decline is part of its beauty. There is time enough to tidy later in the year.

GRASSES

Autumn can be a radiant, glowing time of year in this garden. Rain is the greatest enemy, rain and wind, because everything is becoming increasingly fragile with leaves and flowers hanging on by threads. But if we have a dry few weeks then autumn can drift gracefully into the short days giving us a final fading blaze of colour.

But whatever happens to the autumnal foliage, it is my favourite season for grasses. In autumn, all the rich metallic colours are gathered in their stems and leaves and they can seem to be burnished with gold, silver and bronze in a way that no other group of plants can possibly match.

Grasses have that rare quality in a plant – they sound so good. They catch the breeze as it passes and in sifting it through their leaves, turn it into music. Even in the smallest garden, a single miscanthus is, in its way, as good as a wind chime.

Grasses have been an important part of our garden for quite a few years now and we have made four large beds that were formerly part of the Jewel Garden into Grass Borders. This reflects the fact that the planting is dictated and dominated by grasses and everything else chosen to work with and around – and importantly – through them. But it took me a while to work out how they could be best incorporated and managed within a mixed border containing a wide range of bulbs, perennials, shrubs and annuals. So here are a few grassy thoughts based on my own experience of growing them.

As the borders die back grasses play an increasingly important role and their winter shape, sound and colour are at once the most striking and subtle things in the garden.

I love the stipa family. *Stipa gigantea*, as its name suggests, is big but never overwhelming because the great flowering stems are delicate and do not screen plants growing behind. This means that it can be placed anywhere in a border so is very versatile. Its oaten heads seem to be fired into the air like tracer trails and they catch the midsummer setting sun like burning brands. It likes light and air, so do not plant in a group to bulk it out unless you have the most enormous borders. It also does not like wet, heavy soil, and dislikes being moved once it is established because it has shallow roots and takes a while to recover from the shock of upheaval.

Stipa arundinacea or pheasant's tail grass is a marvellous plant. What makes it special is the leaves, shot with pinks and russets, and the hairy flowers that flop and fall everywhere – two quite different modes. Unlike its gigantic cousin it moves easily and seeds everywhere and the seedlings are easy to pot up, grow on and replant strategically. However, it has a habit of suddenly dying, seemingly for no particular reason.

Stipa tenuissima is delicate, dramatic, and good for a container or the front of the border and has amazing longevity. We grow it in the Dry Garden as well as the Jewel Garden and it is happier in that poorer, better-drained soil. In February, it is one of the brightest things in a border and the feathery heads are silkily irresistible.

The miscanthus family always perform exactly as required: they are tough, and need very little support – which can be a major consideration with grasses in a border which tend to lean all over the rest of your plants. They are drought-tolerant but will also perform well in soggy conditions and invariably have good autumn colour. They are very upright and elegant which makes them good for the middle or back of a border. We grow quite a few and will probably add to this in time. *Miscanthus sinensis* 'Silberfeder' is huge with flowerheads that start out pinkish but take a distinctly silverish turn in autumn. 'Strictus' and 'Zebrinus' both have banded stems although 'Strictus' is more upright and taller. But both are good. 'Malepartus' has flowers that are plum-plumed and open out with a golden thread. It is, as grasses go, almost flawless and will grow in any soil or position. 'Ferner Osten' flowers especially early and has russet flowers. It is of medium height so is adaptable in a border. *Miscanthus* 'Purpurascens' does not really flower with us because it is too cold although, ironically, it is one of the hardiest of all grasses. It has the best orange-bronze foliage of all grasses in autumn but needs moisture to do its best.

At the moment I have two favourites amongst the delicate grasses, although they are so different they are hardly comparable. *Deschampsia cespitosa* 'Golden Dew' is like a mini stipa but its flowerheads gently jangle with gold. It will take much more moisture than a stipa and is completely happy in the Damp Garden. The other choice is perhaps more elegant than delicate. This is a moor grass, *Molinia caerulea* subsp. *arundinacea* 'Windspeil'. It has oaten heads on 2m (6ft) tall, elegant golden stems that gently move with the wind, without – apparently – getting bashed about. Eventually the stems self-prune by snapping off at the end of winter. It is probably happiest in acidic bog but will grow almost anywhere. At the moment we have two young plants and I am looking forward to seeing them grow up.

At this time of year *Carex comans* bronze-leaved blends into the general tawnyness that is overwhelming any border but in spring, when all around it is greening furiously, it can look just plain dead. But I like it as a counterbalance to the surrounding lushness and I like it in winter for playing dead without losing any of its body or form. It does best in full sun.

Some grasses can be too much of a good thing. *Panicum miliaceum* 'Violaceum' (millet) has beautiful flowerheads with seeds hanging like a plum-coloured mane off each stem. It is an annual, easy to grow, move and accommodate but it seeds itself so freely and quickly that it can become an annoying weed. Quaking grass, *Briza maxima*, falls into this category and we have stopped growing it for that reason. Lyme grass or *Leymus arenarius* has fabulous blue foliage and flowerheads like wheat. It is notoriously invasive but too good to ignore. It looks great in the Dry Garden mixed with sedums and *Verbena bonariensis*, but once it gets established in a mixed border it can be the devil of a job to get it out again.

BERRIES

Winter is limbering up around the corner. Soon whole days in the garden will be lost in a sodden grey haze. At such times berries come to the rescue. A berry is the seed-bearing fruit that follows the flower produced earlier in the year. Technically strawberries, blackberries and raspberries are not true berries but grapes, tomatoes and aubergines are. But we all have a pretty good idea of what we mean by a berry in the garden.

So to get the best show of berries you do need to have plenty of flowers and also to allow the flowers to fade and evolve into fruit. This means allowing a certain amount of untidiness in plants like roses with not too much dead-heading. The fruit need summer sun to ripen, so shrubs that flower in May and June tend to make for better berriers than the later-flowering plants. Of course, some roses have more spectacular hips than others and are worth growing for this autumnal bounty, and the early-flowering species roses tend to be best of all.

Rosa moyesii is probably the best-known rose for its bottle-shaped orange hips. It also has wonderful single crimson flowers that speckle the large upright bushes at the back of the Grass Borders in a curiously scattered but deeply satisfying distribution. I have other spectacular rose hips like the great tomato jobs on the rugosas, oval aniseed balls on the dog roses, black ones on the pimpinellifolia and small dangles of orange on *Rosa cantabrigiensis* and *R. willmottiae*.

Hips have an unbreakable link for me with haws, the fruit of the hawthorn, *Crataegus monogyna*. The connection is botanical too, as hawthorn are members of the rose family. Hawthorn is much used in the rural countryside to divide fields, and that is the best place to see the May blossom but it makes a really good garden hedge and birds love it. Of course, a garden hedge that is neatly trimmed before it can flower will pay for its crispness both in flower and fruit, but a standard hawthorn will give you blossom and fruit as well as the freshest green spring leaves there are.

Pyracantha is closely related to the hawthorns and no member of the rose family makes more berries than it. Firethorn flowers are scented with a honey sweetness and bees love them. The more heat and sun that the plant gets the better the berries will be – which is why those trained against a brick wall display better thanks to the reflected warmth. The birds will eat them but not until they have stripped the hawthorns and other softer fruits, and if it is a mild winter they might be left completely alone. I love *Pyracantha* 'Orange Glow' and 'Golden Charmer' and. *P. rogersiana* 'Flava' has yellow berries – although I like my pyracantha to be orange-berried.

Another rosy cousin, and one I feel much more kinship to, is cotoneaster. It comes in many forms from the tiny-leaved *Cotoneaster microphyllus* to the more generous foliage of *C. salicifolius* or *C. serotinus*. All will grow anywhere that the drainage is good, including dry shade, which is why it is so often a component of new-house planting. Bung a cotoneaster in against a shady wall and you will not go far wrong.

Callicarpa berries are downright odd. They are purple, distinctly metallic and grow in clusters along the stems. The shrub is not up to much really – although it has a pleasantly orange autumnal leaf colour – so the berries are the reason for growing it. It needs full sun to perform best and is ideal for growing in a pot so it can be moved into prominence when in berry.

HORSE CHESTNUTS

The horse chestnut, *Aesculus hippocastanum*, is unquestionably one of the finest flowering trees in the northern hemisphere and certainly one of the biggest, reaching a statuesque 36m (120ft) in height and fully 25m (82ft) across.

The flowers, balanced like pyramids of ice cream or white candyfloss, are at their best in May and early June, but the sticky buds break into glowing leaf in early April and the whole tree is a delight all spring and early summer. The performance continues into autumn with the fruit carried in shells studded with rubber spikes inside which are the fabulous seed: conkers.

There is everything to love about a conker. Nothing has a sheen like a freshly exposed conker and not many seeds are such beautiful objects. The name comes from a dialect word for snail shells that were used for the game before the horse chestnut came on the public scene in the early nineteenth century. Before that they were the province of the private park where all gathering of nuts, fruit or wild animals was considered poaching and punished with incredible harshness. So conker trees in public places represent a liberation of nature and public playfulness, as well as decoration.

They will grow fast and in just about any soil, but last nothing like as long as the sweet chestnut with few living for more than 150 years. They were introduced to this country around 1616 and from the first have been valued as a park or garden tree with many a fine avenue planted and treasured ever since.

In recent years many chestnuts have got into trouble with the foliage turning crispy and brown by the end of July. There are a number of causes, from a leaf miner that tunnels along the inside of the leaves to the fungus *Guignardia aesculi* that causes brown leaves, starting with blotches, often with yellow margins. Neither will kill the tree but will sap its reserves of energy. However, horse chestnut bleeding canker, which manifests itself as open wounds on the bark that ooze sap, can kill the tree if it spreads sufficiently. There is no known cure other than painting the wound with a sealant and hoping the tree fights it off itself – which some do.

HOPS

I am writing this from an old hop kiln (which is what we call them in Herefordshire although oast house would be more familiar in far-away Kent). Outside my window the rather bristly lime-yellow leaves of the climber golden hop are visible, spilling from the top of their supporting bean-stick tripod in the Jewel Garden. It is an herbaceous plant that will grow 3–6m (10–20ft) in spring and summer each year before dying back completely in autumn. The young shoots are delicious lightly boiled and eaten with butter like a dish of asparagus.

Wherever hops are grown agriculturally (and alas, this is rarer and rarer) they invariably colonise the hedgerows and reappear for decades, even centuries, thereafter. We have it in our boundary hedge, a ghostly descendent of the hops that were last grown on the neighbouring fields over 100 years ago. We also have a couple of hops in the Jewel Garden, sourced over ten years ago from a local farm that has now fallen under swathes of horrible plastic polytunnels.

The golden hop is the tamed domestic version but it does produce the cone-like female flowers that can be brewed and used as an aid to sleep when stuffed into a pillow. It is a native to this country although hops were not used in brewing until the sixteenth century. Until then, ale was made without hops and, therefore, perished more quickly, and did not have what we regard as the characteristic bitter hoppy flavour. Beer, made with hops, quickly grew in popularity and by 1600 was more popular than ale. The distinctive smell of hops ran through my childhood, as my father brewed his own beer, filling one of my mother's stockings with fresh hops and brewing it on the stove so that the house was rich with their fragrance. The hops in the Jewel Garden are now all that remain.

The lime-green foliage of the golden hop, **Humulus lupulus** *'Aureus', is a striking foil in the Jewel Garden and a reminder that this garden was once surrounded by hop yards.*

The wallflower 'Blood Red' filling the Long Walk. As you cross it it is like plunging into a pool of incredible honeyed fragrance.

WALLFLOWERS

Biennials have adapted to survive the harshness of winter before restarting growth and flowering early in the New Year, so early October is the ideal time to plant them. The ones we use in most quantity here – almost as mass-bedding – are wallflowers.

The basic wallflower has yellow flowers and will seed itself and grow enthusiastically in apparently solid walls and stone – hence the name wallflower. They like good drainage – think of a dry wall – and suffer badly if they sit all winter in wet ground, so I add extra grit to my rich Herefordshire loam. Although they are very hardy, it is a good idea to pinch out the growing tips when you plant them to create bushy plants, and to remove any late, vigorous growth that can be hit by an early frost. It is a mistake to over-feed the soil, which will contribute to a late spurt of growth and, of course, never give wallflowers any kind of fertiliser. The poorer the conditions the longer they will last, and they can keep going happily for more than five years.

Many wallflowers are sterile hybrids that you can only reproduce by cuttings – or buying more seed – but many others seed themselves easily and you can buy seed for about 30 varieties. You do not need a greenhouse to grow them – a line or two in a seed bed in the vegetable garden is fine. Thin them initially to about 2.5cm (1in) or so apart and then again to 15cm (6in), transplanting the thinnings to another line. When I was a child they were as regular a presence in our vegetable garden as carrots or peas.

There are three groups of *Erysimum* varieties. My favourite, *E. cheiri* 'Blood Red', comes from the tallest group that has 12 colours. The Bedder Series is much shorter

and compact with flowers in four hues: bright yellow, primrose, orange and scarlet. You are likely to find these as a random mixture but they can be bought as individual colours, if you shop around. If you are going to grow them in containers, choose the Prince Series that is the shortest of the lot. They come from Japan in five colours, but again, you are likely to buy them as mixed colours. We have *Erysimum* 'Bowles's Mauve' which is a genuine perennial, flowering from March until June, and although not treated as part of a block bedding scheme it is a really good permanent addition to a border.

Although they are easily raised from seed, you can buy them as small, lush plants and they behave like shrubs with a tendency to sprawl, so plant them close together to support each other. Each plant produces flowers sporadically and you need the massed effect to make the most of the colour.

Tulips are the traditional accompaniment for wallflowers and they work well, flowering at around the same time, the tall goblets of tulip rising above the massed foliage of the wallflowers. Daffodils also look good set against the massed foliage, although early daffs will be in flower before the wallflowers. Try the long-flowering tazetta *Narcissus* 'Geranium', which will allow the wallflowers to catch up. Always plant wallflowers first, otherwise half the bulbs will be chopped up by your trowel as you go. There is no rush to get tulips in the ground until November, dotting the bulbs in amongst the wallflowers.

We call wallflowers *Cheiranthus cheiri*, although they are thought to be part of the *Erysimum* family. In any event, they are brassicas, and are liable to get all the problems of the humblest cabbage – so sooner or later they will get a virus, or clubroot, or grey mould or some such disaster. That is their way. But sow fresh seed every year and you will have a healthy new batch ready to take their place.

APPLES

I live in a part of Britain where orchards are still a major feature of the landscape and on a sunny October day the air is wonderfully fragrant with the aroma of ripe apples.

Until the end of 1997 what is now the Orchard was a field occupied by a couple of clumps of field maples and an extremely bad-tempered Shetland pony. But he went (we learnt much later that he served as an alibi for an extra-marital affair – it's a long story) and his field was freed for the orchard I dreamt of making.

Herefordshire is such a richly appley place that I wanted some varieties with local meaning as well as providing good fruit. Of the 39 apple varieties I planted from 1997 to 1998, nine have local roots. They are: 'Herefordshire Beefing'; 'William Crump'; 'Doctor Hare's'; 'Worcester Pearmain'; 'Tydeman's Early Worcester'; 'Stoke Edith Pippin'; 'Tillington Court'; 'Madresfield Court'; and 'Crimson Queening'. All originated from within 20 miles of this garden. They are of this soil. The other 30 include the ubiquitous 'Bramley's Seedling' and 'James Grieve' and the interesting but not local 'Tom Putt' and 'Norfolk Beefing'. (Beefing apples were traditionally used for drying – the fruits, which are low in moisture, were dried in a slow oven and compressed into puck-like tablets before being packed for transportation. Fruits were reconstituted by soaking before cooking.)

I imagined the Orchard as a place where the branches of each tree met above a tall man's head with grass grazed or planted with flowers. It was to be an orchard in its own right rather than a corner where apples were raised. I deliberately chose vigorous rootstocks – mainly MM106 and MM111 – to create half and full standards. The blossom, paths through the Orchard, and light filtering through the branches were as important as the apples themselves.

And so, 14 years on, they are. Of course for the first few years, a lot of imagination was required to get beyond the spindly trees that were slimmer than the stakes supporting them. One of the odd things about planting new apple trees is that although the trees are a whisper of their mature selves, the fruit are full-sized from year one. This is hardly surprising, but of course, it means that they are dramatically large in comparison to the feeble branches that bear them. In the case of a few of my large cooking apples, such as 'Glory of England', 'Newton Wonder' or even the very early 'Arthur Turner', the weight of the fruit can snap the branches. There is nothing for it in this situation but to be tough and remove any fruit that might possibly cause damage until the tree is man enough to bear them safely.

The Orchard rapidly became one of the most used parts of the garden because we have always kept chickens there. The chickens have to be opened up in the morning, fed in the afternoon and closed up at night, so even when I do no gardening I go there three times a day. All walks with the dogs begin there. It is the link from the garden to the countryside beyond, which itself is rich with orchards.

This year, a birthday present of three pigs, housed in the Orchard and showing a great tendency to dig up tree roots, have added gaiety and mess, but in the end it is all about the trees, not the animals. An orchard is a place, and places, even corners

of very small gardens, matter much more than plants. There is a magic to them that transcends the stuff they are made of.

Apples have been grouped into the times that they flower and it is generally accepted that the chances of a regular and good crop of fruit is greatly improved if you grow at least one other from the same group – and ideally a third from a group either just before or just after it. In other words, if you grow a variety like the lovely dessert 'Rosemary Russet' (which I have in the Writing Garden – which, in turn, used to be part of the Orchard) then it is as well to choose another from group 3 to pollinate it. This can be a cooker or eater and there are lots to choose from including well-known varieties like 'Bramley's Seedling' and 'Worcester Pearmain'. But I also have one from group 2, 'Ribston Pippin', to cover the eventuality of an early flowering and others from Group 4, 'Gala' and 'Chivers Delight', which will flower a little later. Suddenly you have an orchard and the likelihood of a decent harvest of apples whatever the spring weather.

You do not need much space to do this. Apples can be pruned and trained to grow in a variety of ways and still produce fruit, from step-over trees with a single branch trained laterally only 30cm (1ft) off the ground, to cordons and espaliers against a fence or trained on wires. Dwarf bushes are usually pruned to an open goblet shape on a short trunk just 60–90cm (2–3ft) high. If you have room for a clematis and a rose in your garden then you have room for a couple of apple trees.

No apple comes true from seed so every tree is grafted onto a rootstock. The choice of rootstock determines the ultimate size and vigour of the apple tree and the top section – almost everything above ground – the variety of fruit. So it is possible to have a 'Newton Wonder' for example, a good cooking apple, growing as a dwarf bush or as a standard with at least 2m (6ft) of clear trunk and a large tree.

Apple rootstocks are identified by letter and number: the most common is M27 for very small bushes, step-over apples and patio trees grown in pots; M9 for step-overs, cordons and small bushes; M26 for slightly more vigorous cordons and larger bushes; MM106 for espaliers, fans, bushes and small trees; and M25 for

I fill a crate beneath each apple tree with windfalls and although these are bruised (so will not store) we cook and freeze them for future use.

full standards growing into proper trees. Smaller rootstocks need better soil and more sheltered sites and are generally less tough.

Apples are usually borne on woody side shoots or spurs, so pruning should encourage these side shoots. The fruit is formed when they are two years old, and some varieties such as 'McIntosh' are better at producing these spurs – and therefore fruit – than others and these are generally better suited to hard pruning as cordons or espaliers.

A very few apples such as 'Worcester Pearmain' produce their fruit on the tips of the new stems rather than spurs, so pruning them back will remove future fruit. Clearly these are not suitable for cordons or espaliers.

However you grow your apples, whether as an orchard big enough to house chickens, pigs and dozens of varieties, or as a line of step-overs along the edge of a path, the fruit should be properly treasured. Early varieties can and must be eaten more or less as they ripen, but later varieties can be easily stored in a cool, dark place. We are out of the habit of storing food because everything is so available all the time. But fruit out of season never tastes as good as properly ripe fruit because in order to travel and store well, it is picked unripe, and the varieties are not chosen for taste but their ability to look good. So do not just grow your own, but store your own.

If you wish to store your apples and have them last as long as possible then windfalls are no use. To keep any fruit it must be perfect when you pick it and handled carefully at every stage of its collection. I have learnt to my cost that you never get away with the odd bruise here, slight knock there. These will always rot faster than the others and spread the decomposition to the fruit stored with them. So pick them very carefully, holding the apple in the palm of your hand and twisting gently so that it comes free in your hand. Then place rather than throw it into a basket. If it does not come away easily then leave it – whatever it looks like.

We have found that storage racks in a shed are ideal. These will have ventilation and humidity as well as keeping cool. In hard frosts a blanket thrown over them is usually enough although in the winter of 2011 all my apples froze solid. I do not wrap mine individually – though my father used squares of newspaper for this – but were you to do so they would undoubtedly keep better. Then, right through to next blossom time, you can regularly withdraw from their lovely cidery ranks to cook or eat fresh, every one a homegrown treat.

Apple care

The ideal situation for any apple is a south-facing slope on rich but well-drained, slightly acidic soil. There should be as little shade as possible but good protection from the wind, especially those from the north and east. Wind is always an underestimated factor on the growth of any plant but apples are especially sensitive. My own orchard has just a few gaps in the hedge on its northern side and three or four trees closest to them are lop-sided as a result.

The best time to plant apples – or any fruit tree – is in its dormant season, which is between October and the end of February. Personally I like to have them in the ground by Christmas. It is now generally recognised that the best way to ensure a good start and long life for the tree is to dig a square hole at least 1m (3ft) wide but no more than one spade, or 23cm (9in), deep. Break up the bottom of the hole, and the sides too if they are compacted, but do not add any manure or compost. Plant the tree in the centre of the hole so that the roots are just covered but the soil firmed well around them. Then stake it diagonally so that the support is facing the prevailing wind. This stake should be removed after three years. Finally water it very well and mulch thickly with manure or compost. Keep the planting hole weed-free and mulched thickly for at least three years, and preferably longer.

My own apples and pears suffer from scab. It is caused by a fungus called *Venturia inaequalis* that is air-borne but overwinters in the scabby cracks and fissures of young stems and on fallen leaves. Chocolate-brown stains appear on the fruit and splotches develop on the leaves that start out olive-green and turn greyish brown. The fruits are often small and misshapen and, in bad cases, will start to crack before they ripen. Secondary infections enter the fruit through these fissures and rot will set in. The leaves often fall early. This means that it is important to cut back any obvious signs of scab on young shoots and to rake up and burn all fallen leaves that might be infected.

Good ventilation will help a great deal. The best way to ensure this is to prune your apple trees so that no two branches are ever closer together than about 30cm (12in). There is no science in this but a lot of common sense. The best advice I had on this was to prune my apple trees 'so that a pigeon could fly through them from any direction or angle'.

Canker manifests itself on the bark of apples (and pears) with the surface sinking inwards. If this encircles the entire stem or branch it then dies back. Often the bark swells and scars at the point of infection. White pustules will show in

The orchard is one of my favourite parts of the garden and its fruit comes almost as a bonus but one we collect and store carefully as winter treasure.

summer and red ones in winter months. Canker is also a fungus, *Nectria galligena*, which spreads on the wind in spring and enters the tree through wounds, pruning marks or infection caused by scab. The only way to treat it is to prune away all signs of infection back into healthy wood. Burn all prunings.

Poor drainage will exacerbate the disease so planning ahead and making sure that the drainage is good when you plant your apple tree will help a great deal. Some varieties are particularly susceptible to canker, and unless you live in a very sunny spot, with free-draining soil, are best avoided. These include three that I have, 'James Grieve', 'Worcester Pearmain' and 'Spartan'. On the other hand, a few show good resistance, including 'Laxton's Superb', 'Newton Wonder', and 'Bramley's Seedling' – which I also grow.

Apple bitter pit shows itself on the fruit with sunken brown spots on the skin, which go through into the flesh. It is more common in large cooking apple varieties or very heavy cropping trees as they demand more calcium from the ground. The best and simplest control is to clear all grass and weeds from around the tree within a 1m (3ft) radius and to mulch each spring with garden compost or well-rotted manure. This will help retain moisture and provide a gentle feed.

QUINCES

Quinces have a delicious fruity, truffly fragrance. It is a scent that reminds you of something ancient and unknowable. Indeed, quinces – that is to say the fruiting tree *Cydonia oblonga*, rather than the ornamental chaenomeles – inspire more reverence than any other fruit. The tree has always been revered as being magically benign. In Jewish mythology the forbidden fruit in the Garden of Eden was not an apple, but quince, the quintessential fruit of good and evil. The Romans considered them an emblem of love and dedicated them to Venus.

Quinces will thrive in damp places that invite trouble for apples and pears and I have four trees in the Damp Garden. I have seriously considered devoting the whole of the Damp Garden to quinces and making it into a kind of side chapel devoted to them. Who knows, when I am looking to simplify things a little, I may yet do that.

Currently I grow 'Leskovac', which is round, prolific and looks like a regal cross between an apricot and an apple; 'Vranja', which is tall, big-leaved and slow to bear fruit (although when it does they are enormous and pear-shaped); 'Lusitanica', which also has pear-shaped fruit, but smaller and the tree is smaller too with a typical sprawl of branches; and 'Champion', which is hard, yellow and apple-shaped.

I pick our quinces at the end of October, as late as the latest apples, letting them cling to the scrubby trees as long as possible. The ripening process is an inexact business at best with the plan only to pick them before they fall and bruise and, like apples or pears, when they come away easily in your hand. They can suffer from blight, their fluff-scuffed skins, covered with a down like the finest hair on a newborn baby's head, blotched like brown ink on blotting paper. The affected parts should be cut out and the rest cooked perfectly satisfactorily. I have grown unflustered by imperfection in these things. The 'success' of a quince can only be measured by the accumulation of the tree, with branches growing like tousled hair, the sugar-pink blossom – perhaps the best of all blossoms of any shade – the leaves that turn exactly the same colour as the golden fruits and then the quinces themselves, blight-blotched or not.

We harvest every scrap, stewing up all the bits of good flesh that can be salvaged even from the most afflicted fruit. Quinces cannot be eaten raw, so cooking them is part of the whole quince experience and we make quince cheese, quince jelly and add a dash of stewed quince to stewed apple to make the sum so much more than the parts. Quince also has a great deal of pectin, so jellies easily and therefore can be stored conveniently, although fewer and fewer people are bothering to make the jams, jellies and cheeses which are the best way to keep fruit beyond a few months. In fact, eating quince in any form has been in decline ever since cane sugar became popularly available and the idea of preserving fruit became a chore rather than a stored-up treat. But perhaps the recent interest in baking and all forms of home-cooking will spur people to grow and process quinces and treat them – as indeed they should be regarded – as one of life's great luxuries.

Quinces are self-fertile – so you can have just one and expect it to fruit – and they are grown on their own rootstock. Pears are almost always grown on quince roots to keep them manageable and from becoming the large, magnificent tree that a pear wants to be. But it seems pointless to try and train or control a quince tree. They are contrary growers, suddenly changing direction or stalling like an old car, so I prune out branches that are rubbing or trailing perversely back down to the ground, but on the whole I let them be and simply admire and celebrate them at every stage, from tree to fruit.

PUMPKINS AND SQUASHES

The weather at the beginning of October is critical for a good pumpkin harvest. These lovely great fruits are the last of the glory boys and the final hurrah of summer. I am a pumpkin fan. They are big, bright and, in my opinion, absolutely delicious as soup, roasted or mashed. However, I have not grown a decent pumpkin or squash since 2006 when they were rather absurdly rampant. That summer was good and hot. That is what the cucurbit family like, lots of heat, lots of water and lots of goodness in the soil. The latter two commodities I can easily supply but heat has been sorely missing, despite the inexorable rise in our average temperature.

All pumpkins are squashes but not all squashes are pumpkins. Pumpkins have hard skins or rinds that mean they can be stored for months if dried in the sun, but many squashes have thinner skins and some, like the summer squashes such as courgettes, never develop a skin thick enough to store.

I sow mine in the greenhouse in the middle of April and plant them out in June, which is quite early enough. I put them in pits that have been half-filled with good garden compost so that the excess soil forms a crater around each plant that will hold water and funnel it down to the roots. They then grow steadily before setting flowers that develop into fruits. Mind you, when you plant them out 2m (6ft) or even 3m (10ft) apart, it seems inconceivable that they will grow and fill the allotted space. The secret is to grow a catch-crop between them, like lettuces or radish, that can be harvested before they disappear beneath a bristly sea of pumpkin leaves, or sweetcorn and climbing beans that will rise above the leafy mulch.

The matter of their greedy spread is a tricky one. Pumpkins take over a large area and in a small garden or one filled with other vegetables, this can be a real limitation. However, this year I have experimented with growing them vertically rather than letting them sprawl. I used tripods made from really stout stakes (coppiced chestnut ordered for the sweet peas but rejected on account of being too thick) that I fixed very firmly into the ground. They have proved effective and I would recommend it as a good way of getting a lot of plant into a little space – but make sure they are secure because even the smallest pumpkin is heavy. In fact we often support the growing fruit with a net that is something between a hammock and a brassiere, lest its weight breaks the stem it hangs from.

Cucurbits (pumpkins, squashes, cucumbers, courgettes, marrows and gourds) all like to be warm from the moment of sowing to harvest. Temperatures below about 12°C (54°F) stop them in their tracks and that is when slugs and snails tuck in, eating the stems where they touch the ground and sometimes gnawing right through them. Once they get growing the plants are robust enough to ignore such nibbling.

By the end of summer the weather is critical. Early frosts will reduce the foliage to blackened rags and cold wet weather will risk the fruits rotting where they sit on wet ground. If it is very wet, place the fruits on a tile or some straw to raise them off damp soil. Ideally they will have as much sun as possible to ripen. The more sun they get the harder the skins will be and the better they will store.

They should be harvested by carefully cutting the stem so that 5cm (2in) remains. This will stop any neck rot. If the rinds are at all shiny it is an indication that they could do with some more sun so dry them outside on a table or, ideally, a drying rack made from chicken wire and posts so the air can circulate. Although they might feel indestructible, treat them gently. Try not to drop or pile up your squashes as they keep better undamaged.

No vegetable is as big and jovial as a pumpkin. Given warmth and water they are easy to grow but almost impossible in a cold summer.

CHILLIES

This, surprisingly late in the year to the uninitiated, is the season of chillies. Given heat and perhaps a little extra light, they will in fact go on producing fruits all year round, and some houseplants – struggling as they are to be perfectly good culinary plants – will obligingly do this. But if you are growing them outside or in an unheated greenhouse, October is the culmination of the chilli season.

Those that love their hot chillies – and that makes up hundreds of millions across the world – do so for their flavour as much as for the sensation of heat. I confess that, late in life, I am coming round to enjoying very hot ones too, although for years I affected to dislike all hot foods. Curries were avoided and Thai taboo. This went on in its anodyne way for about 40 years until I visited Thailand and discovered how subtle chillies could be and how their heat could be finely controlled as part of the overall sensation of any particular dish. So I became a chilli fan and I am certainly learning to love and appreciate the astonishing range of these beautiful and euphoric fruits.

They are easy enough to grow, although a greenhouse will help. They are members of the *Solanacea* family, and if you treat a chilli like a tomato you will not go far wrong, although they are tougher, less prone to disease and respond enthusiastically to unlimited heat.

The seeds should be sown in March or even mid-February if you have a heated greenhouse and kept as warm as possible throughout their life. They are fast to germinate but develop slowly as seedlings – in the case of 'Habaneros', which is what I have grown this past year, very slowly indeed.

I prick them out from the plugs into 8cm (3in) pots and then into their final terracotta pots that are 18cm (7in) wide. This is on the small side and I would use bigger ones if I had them but I bought this batch of pots for a snip and they have done me very well. Over the years I have found that chillies need regular watering as they are growing and setting fruit, but once the fruit has grown they ripen better and faster if they are kept pretty dry.

They need as much heat and sunshine as Britain can give them, so if you grow them outdoors give them a south-facing, sheltered spot. They are best watered every

An unripe sweet pepper, 'Corno di Toro Rosso', growing in the greenhouse. These do very well in surprisingly small terracotta pots.

morning and not after about 5pm as they are prone to fungal diseases and should not go into the evening damp. They also need plenty of ventilation for the same reason. A high-potash feed will make a marked difference to flower and fruit production and I give them a liquid seaweed or comfrey feed once a week.

As a rule, the hotter the pepper is, the longer it will take to mature, but when fully ripe the chilli will have more sugars and therefore a richer, more complex taste and will store better. They will also ripen off the plant: I once picked 99 green 'Cayenne' fruits on a soggy November day that transformed themselves into a deep crimson over a few weeks on the kitchen windowsill and stored for years in a Kilner jar.

CELERY

It is 1972. I am 17 and have endured 10 years of obligatory garden chores imposed on me. I did not like any of it but, despite my reluctance, I have learnt a little. Then, primarily through growing vegetables, I realise that I am starting to enjoy every aspect of growing things. When working outside, hands in the soil, I am complete and growing myself as much as the plants I tend. This was a private, unlikely passion and one that I kept very much to myself.

At about this time I bought the brand new edition of the *RHS Vegetable Garden Displayed*. It became my bible, my bedtime reading, my inspiration, not just in what I did in the garden, but how I did it. Think of an awkward teenager, all over the shop, rebelling against everything and everyone, but absolutely knowing that somehow these men in demob suits and waistcoats, heads turned away from the camera, were telling the truth. I have believed it ever since.

I have that book in front of me now, open at page 81. Celery is being earthed up, a smooth berm of Wisley soil with a Mohican of celery leaves running along the top. It has been an article of faith to recreate that in my own gardens almost ever since.

So far so romantic. But the honest truth is that the resulting celery has nearly always been pretty poor stuff. Wisley has free-draining sandy soil, but for the last 22 years I have been on Herefordshire clay and slugs love the warm, damp earth-

Self-blanching celery (with paler leaves) growing alongside celeriac. Both share the same demand for well-drained but very rich soil and lots of moisture.

wrapped stems. Celery leaf miner often scorched the leaves and added a bitter tang to the stems. Carrot fly took the odd break from my carrots to have a munch and fungal problems blotched the leaves even if they do not, in celery, affect the stems. I have also usually grown self-blanching types too, but argued that as almost all commercial celery was self-blanching it was better and more interesting to grow trench varieties. But the real reason was that earthy berm and the deal I made with myself 40 years ago.

But for the past couple of years, for no particular reason, I have only grown self-blanching varieties. It is a kind of betrayal but an entirely practical and successful one.

Over the years I have always sown my celery seed – which is truly tiny – in a seed tray as thinly as possible and then transplanted the seedlings into plugs. It would probably make more sense to sow directly into plugs but you would have to thin out nine-tenths of the seedlings from each plug and it seems too wasteful. I then grow them on in a coldframe until they are about 10–15cm (4–6in) high, harden them off and plant them out in a grid. I do not water them, feed them or do anything beyond forking in 2.5cm (1in) or so of compost before planting out.

But I think I have the answer to this new self-blanching success. A few years ago I lined a dozen of my vegetable beds with box hedging, which is now thick and mature. My guess is that those low hedges provided the perfect shelter against carrot fly, cold and excess light. It would be a little extreme to proclaim it necessary to grow box hedging for the sole purpose of protecting self-blanching celery but it would be easy to erect a temporary barrier or fence of fleece or perhaps another crop like chives that would do the same job.

I was brought up on celery being served almost invariably as blanched stalks in a water-filled jar but when cooked, celery enriches almost any kind of soup or stew, as well as being a fine vegetable in its own right, the heads gently braised whole. Celery can be stringy – the wisps of tough tissue that lie along the length of the stalk – but the strings are the pathways that carry the nutrients to and from the leaves. This means that so-called stringless varieties are likely to be less robust and smaller than others

Cold

After the experience of the past two winters every gardener should prepare themselves for the risk of frost and cold winds before they do any damage. Until ten years ago you could rely upon a snap of cold weather around the second week of November. Autumn would bare its wintry teeth and all the leaves would clatter off the trees, still frozen, and the ground would be knobbly hard, thawing a little under the midday sun but freezing again as the day drew in. I loved it. It was dry and meant that wheelbarrows could be pushed around without making muddy ruts and the soil could be walked on. But obviously it spelt the end for all tender plants like dahlias, cannas, cosmos and so forth as well as the last of the late summer flowers like heleniums, rudbeckias and asters.

However hardy plants need a dormant period and will be tougher for a cold winter. Many seeds need cold to trigger germination and some plants need a period of cold in winter to trigger flowering. In general, hardy seeds planted in autumn and exposed to cold temperatures produce flowers the following spring or summer much earlier than seeds sown in spring without exposure to cold.

Also hard frost – and I call 'hard' frost an air temperature of below 5°C (41°F) – kills off all the fungal spores and moulds that thrive in wet warmth – the blights, black spots, sooty, grey and white moulds that have prospered so well after a succession of warm, wet summers.

Of course, some plants will suffer, but it is surprising how well plants adapt as long as they have some protection from cold winds and good thick mulch to protect the roots. However, unless you live in a very sheltered, warm spot, my advice is to play safe with tender plants, especially if they are expensive or much-loved. Protecting them is rarely difficult but replacing them can be.

If your garden is in a cold spot try leaving all planting and moving of plants until spring. This will discourage early new growth. Also, leave all dead foliage, stems and branches on plants until spring as they will provide an important insulating layer. Leave your winter pruning – like fruit or roses – until the end of March or even April, so that new shoots develop later. They will soon catch up.

The greatest damage to perennial plants is caused by a combination of wet and cold. You cannot control the weather but you can improve drainage by digging in plenty of organic matter and, if very wet, horticultural grit to keep the soil drier. Add cold winds to the mix and you will have real problems. In this country, that invariably means a north or (worse) an eastern wind. Good

deciduous hedges are the best protection there is against wind, although any barrier like trellis or even netting will make a big difference and help create a series of protected microclimates even within a very small garden.

Any protective layer is effective against light frosts. Wrap tender shrubs like olives or palms in fleece and insulate the ground of your borders with horticultural fleece, newspaper, bracken, straw or a good layer of compost. This stops the surface roots freezing and is particularly important for evergreens. Wrap ceramic pots in a protective layer to stop them cracking in frost.

Cold air drains to the lowest available point. So, do not plant or place tender plants at the lowest point of your garden, even if it is sunny, and be aware that a wall across a slope will trap cold air and create a frost pocket.

Leaf mould

There is no gardening substance so evocative, so clean or sensuous to the touch as leaf mould. It smells deliciously like sunshine on the floor of a wood that is just emerging into spring. It is also enormously useful. Leaf mould (by the way the 'mould' part of that word is significant and I will explain why a little later) makes a perfect element in potting compost, is ideal for mulching all woodland plants and when added to the soil it helps improve its structure. And unlike good compost, which needs turning regularly and has to be made up of a good mixture of ingredients to get the right balance, leaf mould is the easiest thing in the world to make. You just gather deciduous leaves, make sure that they are thoroughly wet, put them to one side and let them quietly get on with the process of decomposition.

There are very few gardens that cannot make some leaf mould from their own trees and shrubs – let alone from the millions upon millions of leaves that fall in the streets outside our front doors. Now you can just leave your leaves to lie where they fall and they will convert to leaf mould where they lie. But in the process they can smother and kill many plants and they do also look untidy. It makes sense to gather up as many as possible to harness their full potential.

When it is half-rotted – after about six months – we use it at Longmeadow as a mulch around the plants of the Spring Garden and also put a layer on the bare soil

As the trees I planted as saplings grow bigger we have ever more fallen leaves and for a few dry, sunny days they make a lovely harvest.

of the vegetable garden and let the worms work it in to make the richest most lovely tilth imaginable. When it is fully made and crumbly rich and smelling of a woodland floor, we sieve it and use it as an integral component of the potting compost. But we do not waste a crumb – or leaf – of it.

My own technique is to use a combination of brush, rake and mower. If the ground is dry enough mowing up leaves is an excellent idea because it chops them as it collects so they rot down faster. In fact, I mow nearly all our leaves, very often gathering them in barrowloads and depositing them on a long brick path in the garden, setting the blades of the mower high, and 'mowing' the path, gathering up all the leaves as I go.

The bulk of my leaves come from the pleached limes and hornbeam hedges, with a good deal of hazel, hawthorn and field maple thrown in. They then go into a large chicken-wire container and are kept wet. The latter is hardly a problem but in a very dry year it means putting the hose onto them every month or so. I stir them about every now and then – more to uncover any dry pockets than anything else – and, hey presto, we have perfect leaf mould by the following spring.

Not every garden is large enough to have a permanent wire bay for leaves. In this case the answer is to put the leaves in a black plastic bag, leaving the top turned but not tied. Make sure the leaves are really wet and punch a few holes in the side of the bag to drain excess water. They will rot down well and can be stored behind a shed or tucked away in any corner.

If you really do not want to store the leaves to make leaf mould, one way of using leaves in a new garden is to spread the fallen leaves over the ground 15cm (6in) deep. Then, using a rotovator (which can be hired by the day) chop them into the topsoil. Unlike green matter, leaves use up very little nitrogen as they rot into the soil, so do not rob the ground of valuable nutrients. This method will create usable topsoil out of the poorest subsoil.

I use well-rotted leaf mould as a component in our homemade potting compost. It is perfect mixed with grit for bulbs like the lilies that we grow in terracotta potsand mixed with sieved loam, garden compost and grit for our normal potting compost.

Garden compost is made by a mixture of bacterial, fungal, invertebrate and insect activity, stimulated by heat and oxygen – which is why you have to turn it. But leaf mould is largely made by fungal activity and is 'cold' insomuch that it does not need heat to spur the fungi into activity.

Fall

There are not many Americanisms that I like, but one that I love is the word 'fall' for autumn. In every way it perfectly suits the season with its falling leaves. The degree of colour change in fall is dependent upon late summer weather, when hot days and cold nights stimulate the production of chemicals closely related to carbohydrates that produce red pigmentation. The leaves convert starch to sugar to feed the tree but cold nights stop it moving from the leaf back to the roots. This accumulation of sugar in the leaves often results in red pigmentation and as the green chlorophyll begins decomposing as the days shorten so the red comes to the fore. The greater the difference in temperature between day and night – in other words the hotter the days between late July and early August are – the more extreme the leaf colouration will be.

Yellow leaves are coloured by a different process – essentially the removal of chlorophyll, which produces green pigmentation, to reveal the yellow that is there all the time. The yellowest of all autumnal trees are those of the English elm, which you will only see in a juvenile version or in a hedgerow, as since 1975 all mature trees have been killed off and young ones succumb to Dutch Elm disease at around 6m (20ft) or 15 years old.

Leaves fall when cells break down in the abscission layer between leaf stalk and twig. A corky scar forms over the wound that this causes, protecting the tree from infection. Some trees cannot form this scar-tissue so they do not drop their dead leaves until the new ones are ready to push them off the following spring, which is why beech and hornbeam keep their russet leaves all winter. Poplar, birch and willow fall early but oak can hang on well into December, although I never cease to be surprised by how arbitrary the actual moment of fall can be – the leaves often tumbling down with a clatter after a heavy frost on a seemingly windless day.

Evergreens do not change colour but very few hang on to their leaves for more than a year. In practice, 'evergreen' means that the leaves can overwinter before being renewed in spring. However, although they do eventually fall to the ground, do not add evergreens to your leaf mould pile, as they take much longer to break down.

October jobs

Here are some of the jobs that I try to do before the end of October. But note that all-important word 'try'. The beauty of this time of year is that most of these jobs can take place at any time over winter. This list is for those with the time, energy and enthusiasm to make the most of the fading light.

Start digging any ground that you want to replant this winter or use next spring. Doing it at this time of year means that it is accessible, dry and there is more daylight to do it in! But if this seems daunting remember that an elephant can only be eaten one mouthful at a time. Nibble at it. If you do 30 minutes a week in two 15-minute sessions it is surprising how much you can get dug by Christmas. Do not try and break it down but leave the soil in large slabs for the weather to break down over winter. Mulch raised beds – if you don't have any, October is an ideal month for making them – with 2.5cm (1in) or so of garden compost as they become clear, leaving the worms to work it in ready for sowing or planting next spring.

If you do not already posses them, invest in horticultural fleece and some cloches. These are only useful if you employ them before you need them. Cloches are good for rows of vegetables, keeping them dry as well as warm (although I always leave the ends open – happy to trade some heat for some ventilation), and fleece is the best temporary protection against frost, either laid over small plants or draped over shrubs and bushes.

Keep dead-heading throughout October, particularly the short-day plants like dahlias. This will extend their flowering season and squeeze the last bloom from them.

Save yourself a fortune by collecting seeds from perennial plants, using paper (not polythene) bags. Always label seed packets immediately. Store in a cool, dry place until ready for sowing.

It is not too late to take cuttings and there is no more rewarding success. Choose healthy non-flowering growth, use a sharp knife and very free-draining compost (I use 50:50 sharp sand and sieved leaf mould) and keep the humidity high. Most things will strike now and overwinter successfully without needing potting on.

Salvia elegans *flowers very late in the year and yet will not survive frost so it tends to adorn the greenhouse, where it gets the protection it needs, rather than the garden.*

You can plant or move deciduous trees, shrubs and hedges even if they are still in leaf as they have finished growing and the soil is still warm so the roots will begin to grow immediately. I once moved a four-year-old, 20-metre long hornbeam hedge in October. It never batted an eyelid and grew away the following spring stronger than ever. It is essential, of course, to give them a really good soak when you do so and to repeat this weekly until the ground is really wet or the leaves have fallen. But, if you are planting or moving a number of trees or shrubs it is best to prioritise the evergreens as they need to maximise root growth before winter kicks in.

Plant or move biennials such as forget-me-nots, wallflowers, foxgloves, onopordums and mulleins. Dig up healthy *Verbena bonariensis*, cut backand pot up to use to take cuttings next spring and take cuttings of penstemons and salvias.

Continue planting spring bulbs but wait another month for tulips.

Cut back and compost all rotting foliage in the borders but leave as much winter structure as possible.

Unless the weather is bad, most leaves do not start falling until November but gather them all and store every last one – nothing makes for a better soil conditioner or potting medium. If you do not have somewhere to store them sort this out early in the month. A simple bay made from four posts and chicken wire is ideal.

Sow 'Aquadulce' broad beans and an early garlic variety like 'Sprint' or 'Thermidrome'. Sow sweet peas in pots and overwinter in a coldframe.

Keep cutting the grass for as long as it keeps growing, however, it is better to have the grass too long than too short over the winter months. Rake out thatch and moss and add to the compost heap.

Cut off any hellebore leaves that are obviously diseased and mulch around spring-flowering perennials with a 50:50 mix of last year's leaf mould and garden compost.

Prune climbing roses by removing old stems and any damaged or crossing new growth and cutting all lateral shoots back to a healthy leaf bud.

Tie all climbers up securely.

We give our hornbeam, box and hawthorn hedges a light trim in October which keeps them crisp right through the winter and looks really good when everything else has sunk into decline.

Finally, get outside and relish every second of sun. It will be a long time gone.

Dead-heading dahlias as soon as the flowers are spent will increase the quantity, quality and duration of their flowering period in autumn.

NOVEMBER

November is an odd month in the garden, always a bit better than I had remembered and a bit worse than I had hoped for. The truth is that the garden spirals out of control, running down the year like water emptying from a basin. It grows sunken, and huddles into itself. The clocks have shifted and although I love the way that dawn returns an hour earlier, the price is an end to any kind of evening outside. It is unquestionably the worst month for garden and gardener alike.

There is nothing you can do about this, so common sense and experience suggest we just go with it. The trouble is that it is the antithesis of everything that we – I, at least – love about the garden. I love the light and growth and sense of hope that leaps out from the most modest flower. I love the way that it sets the world to rights – even in the teeth of almost impossible odds. November does not do this, poor benighted month that it is.

We can still be busy. There is ground to dig, deciduous hedges, trees and shrubs to be planted, tulips to be got into the ground, and pots and borders to be cleared and put to bed for winter. Given the shortness of the days and the vagaries of the weather, even in this climate challenged era, there is more than enough to do in the time available. But all my inclinations are to take stock and to consider and plan rather than act. It is a chance to get to know yourself and your garden a little better.

The first step is to spend time just getting to know your plot. I am a great believer in walking every inch of the garden at least once a day, whatever the season, and whatever the weather. If I cannot do this in daylight then I do it at night by torchlight, as often as not in November in blustery rain. No matter. It is all part of the garden and that knowledge filters into the process of making it look wonderful in the high days of summer. So it is worth knowing where the light falls at 3.30pm on a bright mid-November afternoon, or where the east wind just catches the corner of the border once the leaves have fallen. In my own garden there is a path between the Jewel Garden and the Grass Borders – just a metre or so – that is crunchy underfoot with frost before the ground around it. Why? I don't really know, but I do know that the plants that flank that metre or so of path will be the first to be stricken by frost and therefore need to be planted with that in mind.

WINTER POTS

I am a sucker for large individual pots. The first really good one that I bought was more than 30 years ago when Sarah and I were spending Christmas in the South of France. We had seen some lovely pots and asked an English ex-pat couple where we might buy one. We were instructed to 'bat on down to Biot' where we would see fields full of them. So we duly batted along the coast road and sure enough we discovered acres of lovely jars. We chose our favourite, just fitted it into the car, securing it with a seatbelt, and drove it home to our Hackney garden. It was a beautifully curvaceous oil jar that was put at the centre of the garden, although never planted it up. Then in the hurricane of 1987 it was blown over and smashed. We repaired it as best we could and it adorned our garden here in Herefordshire for a further ten years – always unplanted – until one of the cats arched its back against it and toppled it over, smashing again.

We gathered the bits and took them to a local potter to see if it could be mended and also for him to make a copy. Both projects eluded him because he said that it was more than 500 years old and so delicate that, on scale, the clay was thinner than an eggshell and it defied all the rules of pottery. In the end it remains in pieces but we have a much thicker, cruder copy in the Jewel Garden. Now this is rather a long story to show how one can become attached to pots as objects – and we have dozens that we love – but I have not yet worked out a good way of using them for plants. Nothing looks quite right or at least better than the unplanted, unadorned pot.

This pot is a copy we had made of a very old pot we bought more than 30 years ago in France which got smashed in the great storm of 1987.

Really big, dramatic containers planted so that they are like a huge, vibrant bunch of flowers are always a Good Thing, but they are genuinely difficult to do without becoming messy or at least only successful from certain angles or for a short period of time.

But failing such regal displays I am now inspired to use repeated, identical pots, identically planted rather than one big display. At this time of year you can create an instant effect by using simple topiary shapes from box or holly, or grasses such as *Festuca glauca*, *Briza media*, or the brown *Carex comans*. If you are prepared to splash out, a pair of citrus plants in pots makes a superb display flanking a doorway or entrance, although the plants will need protection from all but the lightest frosts. However, this can be done on an ad hoc basis simply by draping fleece over the plants and removing it as the temperature rises.

I am planning to plant a series of evergreen ferns in pots for the shady courtyard that is the antechamber to the Walled Garden. I am not after anything exotic. Hart's tongue (*Asplenium scolopendrium*), the soft shield fern (*Polystichum setiferum*) and the Japanese shield fern (*Dryopteris erythrosora*) are all lovely and will thrive in the cool shade and add a depth to what can be a very bleak area in winter.

Early November is not too late to plant up some pots with bulbs. The conventional advice is to plant these in layers, starting with the largest such as daffodils and tulips at the bottom and working towards the top through hyacinths, muscari and crocus. This works well if you only have one or two larger pots but I would suggest planting a single type of bulb in as many small pots as you can get hold of. The only important thing to remember is that all bulbs other than fritillaries and snowdrops like very good drainage, so mix any potting compost you buy with an equal volume of grit. Snowdrops look lovely in small pots but are best grown from plants rather than bulbs, so wait until February and buy or plead for some plants 'in the green' – ie during or just after they have flowered. Pot them up and leave them in a shady corner before bringing them out after Christmas. However, whatever you pot up and whatever you pot them into, make it simple and repeat it!

The backlit flowerheads of miscanthus shine with a fluffy incandescence in the low autumnal sun.

SALVIA ELEGANS

Salvia elegans is known as the pineapple sage because the leaves smell strongly of pineapple when crushed between your fingers. For most of the year it is without flower, then, astonishingly late in the season, it produces scarlet flowers completely at odds with the surrounding gloom and waning of the year. It is tender, which means that it must either be grown in a pot and brought under cover once it gets frosty, or lifted after the first proper frost and put in a pot over winter in a coldframe before replanting in June. If you have a conservatory or generous windowsill it makes a very good houseplant, although it needs very little protection and is perfectly happy as long as the temperature remains above freezing.

It comes from Mexico and flowers in response to the shortening days rather than the actual season, so like poinsettia, I suppose that it could be tricked into believing that it was closer to the Equator by limiting its access to light. But it is such a handsome plant that I do not like to muck about with it and simply hope that the frosts do not come before the flowers have a chance to perform.

It takes very easily from cuttings. I use the side shoots that grow at 45 degrees between stem and leaf, putting half a dozen to a pot, and leave them in the greenhouse or a coldframe all winter. I have learnt to resist watering them much, just leaving them pretty dry and frost-free until the end of May before potting on or planting directly outside.

MISCANTHUS SINENSIS 'ZEBRINUS'

Miscanthus sinensis 'Zebrinus' is a lot of name for a lot of plant both in volume and impact. This miscanthus develops horizontal banding which, because every blade of grass lilts and leans, only hints at uniformity. By autumn the end of each leaf spirals and corkscrews away from the main body in a kind of final abandoned flourish. The zebra striping in the leaves is controlled by temperature and turns from plain green to banded as the season gets hotter. It is ideal for the back or middle of a border and after a year or two to establish – most grasses are slow to get going – it grows to about 1.5m (5ft) tall and will resist almost any kind of weather. From the second year it produces purple panicles of flowers at the end of summer.

All miscanthus like a rich soil in full sun but are very adaptable and will tolerate most soils and positions. The leaves should be supported and left to stand all winter where, amongst their visual qualities, they will add a musical rustle in the faintest breeze. Cut or pull them back in spring when you see the new shoots appear.

M. s. 'Hinjo' is a smaller, slightly fuller version which spills down more readily upon itself and is therefore better for the front of a border.

MAHONIAS

Mahonias, although not glamorous, and never the star of any garden, are superb for lifting winter gloom with their bright racemes of blazing yellow flower which, at their best, smell as good as lily of the valley. Their blue berries can be made into jam, too.

When I was growing up in Hampshire they grew in the deep shade of the house, etiolated through lack of light, unpruned and, to be honest, not much-loved. They were there because they had always been there. Being accommodating plants they soldiered on in the shade and chalky soil, modestly doing their thing. I did not consider planting any mahonias in my first, London garden, but when I moved to the country in the 1980s we inherited a dozen large shrubs, planted between huge redwoods, which lit the winter like gentle, fragrant lamps. I now have a couple of plants in the shadiest corner of the Spring Garden.

Mahonias get their name from an Irishman, Bernard McMahon, who emigrated to Philadelphia in the eighteenth century and who set up a nursery that became a famous horticultural meeting place. He also wrote the first American gardening book, *The American Gardening Calendar.*

The first to flower is *M.* × *media* 'Lionel Fortescue' which will draw all the bees to its upright lemon flowers. *M.* × *media* 'Charity' is probably the best-known mahonia and will grow to a substantial (and very prickly) shrub, 2 × 2m (6 × 6ft), given space and time. The primrose-yellow flowers are strongly scented of lily of the valley. Another cultivar of the media hybrids is 'Winter Sun' which is smaller than either 'Charity' or 'Lionel Fortescue' and has deliciously scented flowers. *Mahonia* × *media* is the result of a cross between *M. lomariifolia* and *M. japonica*. The latter is very hardy, tolerating both extreme cold and shade although it has a tendency to become rather lanky and gaunt in old age and needs pruning to hold a good shape. However, it is a robust, good plant and has wonderfully fragrant flowers. *M. lamariifolia* comes from Burma and Yunnan in China where it can

become a really big upright shrub, reaching 12m (40ft) tall. Its flowers are carried in a tuft, it has really good foliage but it is much more tender than *M. japonica,* it is not hardy below -10°C (14°F) and unhappy in any windy, exposed spot in winter. I certainly could not grow it in my garden here on the Welsh Borders.

One of my plant-hunting forebears, George Don, brought back *M. nepaulensis* from his travels in Sikkim, which, like *M. lomariifolia*, will reach 9–12m (30–40ft). I suppose I ought to grow it in my garden out of family loyalty but it is also rather tender.

A better-known plant-hunter was David Douglas, famous for introducing the Douglas fir, and who added more than 200 new species to our gardens and landscape. Like my own plant-hunting forebears, Douglas was a Scot born near Perth. At the age of 11 he began a gardening apprenticeship at the Botanic Gardens in Glasgow and came under the mentorship of the great Sir Joseph Hooker (who later became the first Director of Kew). In 1823, Douglas was sent to the east coast of America to hunt for new plants by the Horticultural Society of London (which became the Royal Horticultural Society). This was a great success and on his return he was then sent to the north west coast – which involved an eight and a half month sea journey right round the Cape Horn to get there. Very early on in his three-year trip, packed with adventures that make Indiana Jones look a wimp, he discovered *Mahonia aquifolium* or the Oregon grape.

The American mahonias are generally shorter and more spreading than the taller Asiatic ones. American mahonias do better in sunny sites with well-drained soil whereas the Asian ones like some shade and, as in my garden, cope happily with heavy soil. But do not let this put you off attempting to grow any mahonia anywhere – most will adapt perfectly well to wherever you place them.

Mahonia is incredibly useful for brightening dark, dry areas where little else will grow. This **Mahonia x media** *is in the shadiest part of the Spring Garden.*

Although it is underplanted with spring bulbs and wallflowers, the Long Walk is deliberately planted to retain a strong green spine down the middle of the garden all year round.

EVERGREENS

In winter, when there is more space than at any other time of year, the gardening of emptiness is at its hardest. Anyone can make a green path flanked by immaculate tall hedges look good if they are in the full flush of midsummer health, but a grey November day is less forgiving. The eye does not wash over rough grass, straggly branches and floppy evergreens, but sticks at each irritant.

Green is the answer. Green and still yet more green. Green plays too many structural and balancing roles to be just colour. It is the white page that the rest of the garden is written upon. Much is written about the 'bones' of a garden provided by hedges, trees and clipped shrubs, but maybe we underestimate the sheer colour of evergreen plants, be they in the form of a box hedge, large tree or leaves of a border plant such as a hellebore. It is a reassurance and comfort that life is ticking over and will return with gusto in spring. Of course, flowers are wonderful at any time of year but by now they are desperately isolated. All that are left are survivors and stragglers, hanging on into the wrong season like the last to leave a party. At no other time of year is there such a tangible, measured absence. I suppose that this is the pendulum that balances the astonishment of every spring. However you justify the need for this dormancy in the greater scheme of things, the truth is that the garden lies exposed and you need all the structural green you can get.

The effect can be achieved simply by repeating a favoured evergreen plant. The most obvious outcome of this approach is the box hedge, defining and lining all the spaces that were filled with flowers. It is a simple enough idea but has been practised for centuries across many cultures and one that I have assiduously pursued, planting box hedging to frame nearly every border in my own garden. Having been to all intents and purposes invisible for half the year, these hedges, now free from the competition of the flowers that they enclose, suddenly leap out and become, over winter, the strongest and arguably the best thing in the garden.

The secret of making the garden look good in winter is to have it tightly trimmed, so that by the time the leaves fall the remaining structure looks sharp and distinct – whether that's a 6m (20ft) tall hedge or swirling parterre, or some form of topiary from a simple box cone to a pack of hounds running in full cry, clipped out of yew (as one of my friends has).

But, this is not just a hedge thing. A single green plant in the dead of winter will do the trick if planted judiciously. Hebe, choisya, Portuguese laurels, holly, Irish yews, mahonia, yuccas, viburnum, camellias, skimmia, pyracantha, euonymus, rhododendrons (some), trachycarpus, hellebores and, absolutely best of all, the holm oak – the point of the list is that it is eclectic and unthemed save for the undimmed greenness of their leaves.

Our only indigenous British evergreens are holly, ivy, yew and juniper. All others are introductions. We now think of evergreens as being tough but the nomenclature of a greenhouse as being synonymous with a hothouse shows how 'greens' were considered exotic and not hardy well into the eighteenth century. In fact, many evergreens are tough enough to survive winter under a protective layer of snow or even ice. Most have leaves with a thick skin, a simple shape and a waxy surface – all designed to keep moisture in the leaf for as long as possible. Wind, frozen ground and even sun are their chief enemies as the plants are most prone to dehydration when the green leaves are losing water faster than the roots can extract it from the frozen earth. I have lost large box plants on a roof in London through desiccation caused by winter winds whilst identical ones 3.3m (12ft) below survived the same weather without a damaged leaf. A strong wind on a chilly day will do more damage than any amount of snow or ice. It always astonishes me when I see huge bay trees growing in London whereas in my exposed garden they must be given the same degree of protection as my oranges and spend their winters, suitably enough, in the greenhouse.

Very few evergreens keep the same leaf for more than a single year, but instead of shedding their leaves in autumn they hang on to them until new ones start to grow in spring. This means that, given mild weather, they are able to continue growing all year but need a steady supply of water even when not only is there no supply of atmospheric water but also any moisture in the soil is frozen and therefore inaccessible to the roots.

SWEDE

It seems that the world is divided into those who love mashed swede and those who cannot abide it. The Don household is firmly in the camp that adores it. In my early twenties Sarah and I spent a winter on the North York moors and I helped my neighbouring farmer with odd jobs and he paid me for these with milk, potatoes – and swedes, although the latter had to be dug from the field where they were grown as winter grazing for sheep. They were very good.

In the North it is often confusingly called turnip (thus haggis and neeps) and the Americans call it rutabaga which comes from the Swedish 'rotbagga'. But it is all the same swede, *Brassica napus*. The name gives away that it is a member of the brassica family and should be grown as part of the same rotation of cabbages, radishes, mizuna and kohl rabi.

Swedes did not reach Britain until the mid-eighteenth century via Holland. They were originally grown as animal fodder and eaten only as a last resort by the rural poor, who discovered that they could be delicious. The name is an abbreviation of 'Swedish turnip' and in fact, it is thought to be a cross between a cabbage and a turnip. Like turnips, they are biennial, making the root and leaves in the first year and flowers and seed in the following spring.

The seeds are sown in spring and allowed to grow slowly through summer, with the best harvest in autumn and early winter. Like turnips, they need early thinning and should have lots of room – 23cm (9in) is a good rule of thumb. Other than that they are remarkably trouble-free, although flea beetle will eat tiny holes in the leaves of young seedlings which can slow growth, and like all brassicas, they are susceptible to clubroot, but this is only likely to be a problem on poorly drained, acidic soils. Digging in plenty of compost and liming the bed before sowing will help avoid this.

RADICCHIO

At this time of year salads from the garden are a precious treat – and one that I try to enjoy every day, regardless of the weather. This takes some organising and judicious choice of the right leafy plants, which is where radicchio is invaluable and completely delicious. It is beautiful too adding a deep carmine tone to the salad bowl.

Radicchio is a confusing term because it actually simply means chicory, but for us it refers to those that share the same characteristic of having deep red leaves. I regularly grow 'Rossa di Treviso', 'Rossa di Verona' (see also pages 336 and 337 for photos) and 'Castelfranco' and all of them start life with pure green foliage growing luxuriantly all summer. As the weather cools this is replaced by much smaller, rosetted red leaves, and the green leaves, which exist to help establish the deep taproot, can be cut off and removed to the compost heap by the barrowload, their job done. This taproot will sprout two or three new flushes of red leaves between autumn and spring if they are harvested by cutting close to the ground.

The seed is best sown in May, either directly into warm soil or seed trays, and then pricked out and grown on in plugs. The final spacing in the ground should be a generous 30cm (12in) to allow the foliage to spread. Watch out for 'capping' as radicchio has a tendency for the outer leaves to become brown and slimy if too cold and wet. Peel away this nasty layer to reveal perfectly healthy red growth beneath it. I find open-ended cloches the best way to counter this.

PARSNIPS

Parsnips have grown in this country since the Romans introduced them, and were widely eaten before the potato became the staple starchy vegetable. They taste much better after a frost has intensified the sugars in them and will withstand any amount of foul weather, remaining unharmed in the ground until spring or until you want to eat them. If left until spring they will start to use the energy in their roots to produce a flowering stem.

They are perfect baked or roasted with a joint of meat and I love parsnip purée. You simply boil the roots and then liquidise them with some of the cooking water and lots of butter or cream.

They are not at all difficult to grow but the large flat seeds are slow to germinate and need a long growing season to mature. Sow seeds in March or April on a piece of ground that will not be needed for a year. The soil, obviously, must be dug to a good depth if the roots are to develop to any size but do not add any garden compost or manure as this may cause the parsnips to fork or split – and will also encourage a lot of lush foliage above ground at the expense of the roots.

Digging parsnips is nearly always done in winter cold or wet but a tap will quickly wash them clean and the intensity of taste is worth the trouble.

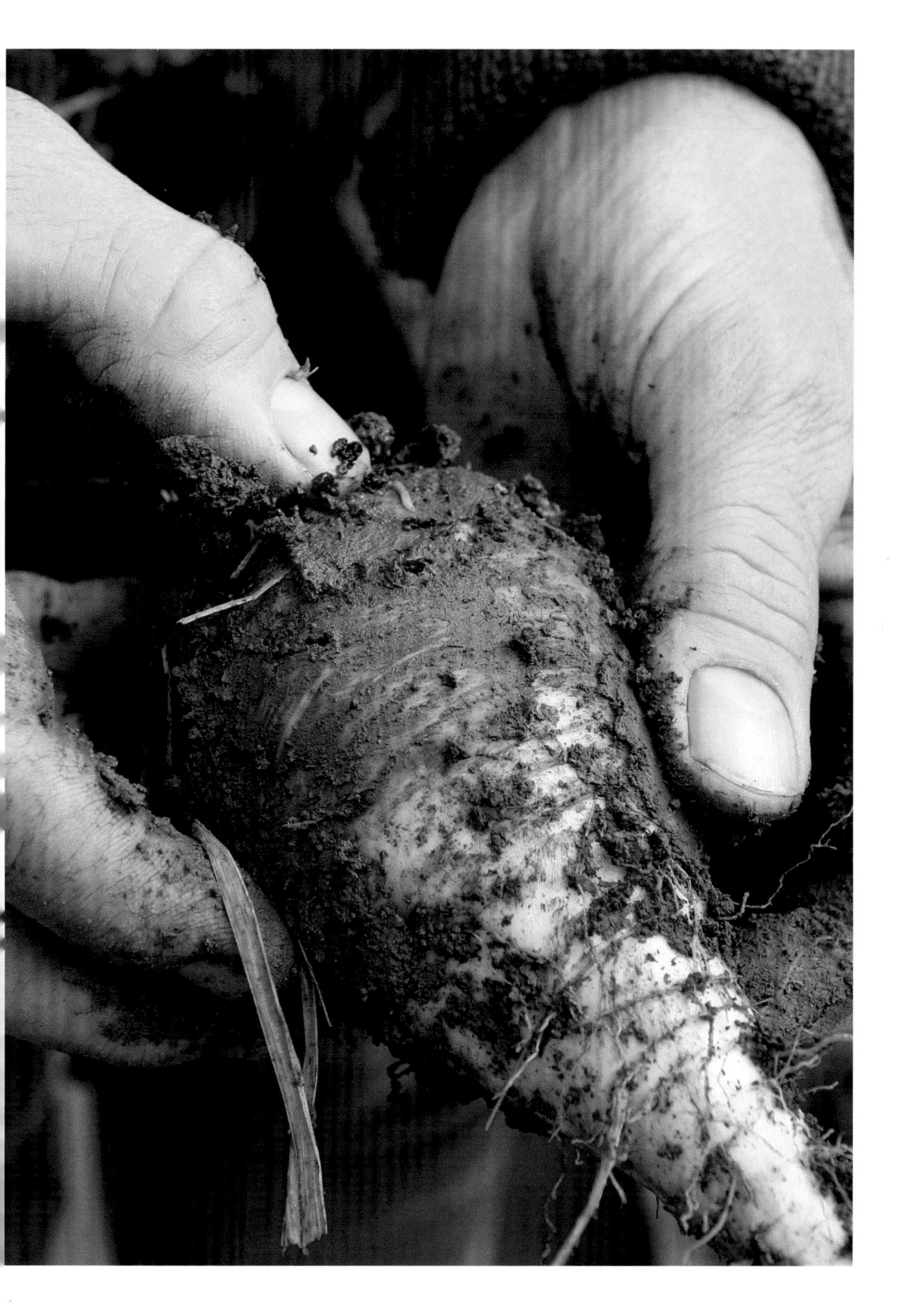

I always add a sprinkling of radish seed in the drill at the same time as the parsnip seeds. The radish germinates quickly, marking the row and helps space the parsnips out. They then get pulled and eaten before the 'host' crop gets properly underway.

LAMB'S LETTUCE

Lamb's lettuce (*Valerianella locusta*, also known as corn salad, and in America as maché) comes into its own now as a delicious component of a winter salad. It is a native plant which is hardy but at its best in winter months if protected by cloches or an unheated greenhouse. In fact it can be grown at any time of the year but it has a tendency to run to seed in warm weather so is best kept as a cool-weather crop.

The small oval leaves are very tender and just slightly mucilaginous with a distinctly nutty flavour. The plants that I harvest in November through to February are sown in August, and although they germinate fast, they grow very slowly throughout the autumn, putting most of their energy into developing a fibrous root system before producing a good crop of edible leaves. Then in January I sow another batch that will be harvested between the end of March and June.

The best way to harvest them is to cut the whole plant with its mass of leaves at ground level and you should get a fresh crop within about four weeks. My favourite varieties are 'Verte de Cambrai' and 'Verte d'Etampes' which tend to be a little hardier than others.

CABBAGE

You can dress up asparagus, fresh peas in the pod, even a bunch of carrots as being sexy, wholesome and beautiful but the average cabbage is not cut from that same cloth. They are not glamorous. However, I love them, both to eat and to look at and, I confess, as part of the process of growing. They are satisfying at every level. They are also one of the oldest and most staple vegetables known to man, across all cultures and civilisations.

Cabbages as we now know them have been bred rather than occurring naturally. The cabbages of the ancient Egyptians and Greeks were loose-leaved and the first headed cabbages appeared some time in the first century BC. The head is an enlarged terminal bud, which started out something like a Brussels sprout but was gradually bred to be larger, and the familiar 'drumhead' was developed in north Europe in medieval times although well into the sixteenth century most British cabbages were loosehead.

Red cabbage was one of the first variants, followed by heads ranging from white to a deep green with varying degrees of looseness. To put that into context, savoys, which I think of as having a firm head, are classified as loose-leaf. Brassicas all tend to hybridise but few seedlings come 'true'. Therefore, distinct cabbage types are the result of careful breeding by centuries of cabbage-lovers.

I grow four types of cabbage, savoys, drumheads, red and spring. Savoys are my favourite. They originated in Italy, probably from the old Roman types, and have a sweet butteryness that is surprisingly delicate. They also look fantastic with their convoluted, brain-like folds that crinkle across the range and depth of their greens. Winter whites or drumheads develop a solid white head and were developed primarily for sauerkraut. They mature in November and December although store exceptionally well – which is why they are so often available. 'January King' is an old French variety with a characteristic pink shading to its blueish leaves. Red cabbages are the toughest of the lot and store very well. They are also the handsomest things in a winter vegetable garden and worth growing for pure decoration.

Spring cabbages are either loose-leaf (spring greens) or make a heart – often pointed – but both types mature very slowly over winter from a late summer sowing. Summer cabbages tend to be pest-free because they miss the butterfly attack of summer and are consequentially perhaps the easiest cabbage to grow. I sow mine in March, plant them out in May and harvest them from July into autumn.

If the young plants are hardened off properly and are growing well, then I find that slugs and snails are not a problem. In fact, until late summer they are not a problem anywhere in my garden. Clubroot is not something I have ever had to contend with. In fact, my organic garden is largely a trouble-free heaven. It is therefore fitting that my biggest problem should be butterflies.

Every year the large and small white butterflies waft delicately into my vegetable garden and do their worst with my assorted brassicas. The large white lays its eggs on the underside of the leaves from which emerge yellow and black caterpillars. The small white lays deeper into the plant and its green, perfectly camouflaged caterpillars do their work less conspicuously but to just as noxious effect. The best cure is prevention – covering the plants with a fine net from the minute they are planted until late September to

stop the butterflies from laying their eggs. However, I do not like nets – they look ugly and tangle and have to be lifted every time you weed or harvest. So I go through the plants every day, gently crushing caterpillars. Not pretty but effective enough, satisfying even, in a revengeful way.

However, cabbages are tough things. Experience shows that it is rarely too late to clean off the caterpillars – and they will go on hatching well into autumn – and let the plant have another go at growing. A cabbage after all is a loose-leaved thing that has been bred to have a heart. It wants to develop new leaves be it red, kale, savoy, sprout or broccoli. This is where the good soil comes in. In rich, well-manured, dug, unacidic soil it will grow and grow, despite the attention of hundreds of butterflies and their hungry offspring.

Cabbage growing with exuberance in the new Top Veg Garden. I find that cabbages of all kinds respond to a good dressing of compost before planting. It is a shame that cabbages are tainted by association with over-boiled institutional food as they are handsome plants that should make delicious eating.

Bonfires

Longmeadow has a bonfire in the same way that it has a compost heap or a vegetable garden. It is not just a function but also a place. We talk about 'going to the Bonfire' in the same way that we go to the Orchard. It is also a place that steadily gets bigger and bigger too, with the heap of unburnt material accumulating for weeks at a time and the ash beneath it also piling up. Every year I sieve barrowloads of it and put the fine grey ash onto the garden, but just as hedges always get bigger however much you trim them, the bonfire heap slowly accrues and accrues until it becomes a crusted grey volcanic lump.

Bonfires are not seen as a necessary part of the garden today although when I was a child every garden had one – albeit an incinerator. Nowadays bonfires are not deemed environmentally friendly – but like everything it is a question of weighing up the options and using common sense.

You should certainly avoid burning anything that can be composted, including most packaging and soft prunings. You should not burn anything that can be easily recycled, like newspapers. Most dry stalks and non-thorny woody prunings can be shredded and a shredder is an excellent investment. My own experience is that you should go for the most powerful one you can afford so that the widest range of material is composted.

One word of warning for dealing with a pile of garden waste, whether it be for shredding or burning, just be mindful of the excellent winter cover it provides for hedgehogs, grass snakes and toads in particular. We tend to light it frequently so the heap never gets too big or old and to shred in spring – after the hibernation period is over.

I burn all our really nasty weeds such as couch grass, ground elder and, nastiest of the lot, bindweed. Wood that cannot be burnt on the fires indoors is also chucked on the fire, especially to get it going.

The alternative would be to bag it all up, drive it to the local tip and have it taken from there to be buried. I have no idea what the actual carbon emission equation would be but I cannot believe that an occasional bonfire comes out significantly worse than all that driving around.

The secret of burning such weedy material – which inevitably has soil clinging to it and a lot of greenery – is to stack it up by the bonfire and add it slowly, little by little, so that the fire is no more than a trickle of smoke struggling skyward.

Birds in the garden

I have kept chickens all my life and apart from the wonderful flavour of the steady supply of fresh eggs I value them highly as a useful and pretty addition to the garden. They spend their day scratching around the Orchard, eating larvae and grubs and keeping my fruit trees largely pest-free. I even kept a couple of hens that roosted in the outside loo when I was a student and certainly a small hutch for a couple of hens can be moved around a lawn without causing any damage to the garden. But you do not have to keep hens to enjoy birds in your back garden and as I get older I enjoy the presence and company of all birdlife in my garden more and more.

Every garden has some birds in it but some have more than others, and the longer that I garden the more I am aware that the number of birds in your garden is as good a measure of its health as anything else. If a garden can attract and support lots of birdlife it must also be rich in the insects and seeds that they need to sustain them, which in turn implies a rich and varied ecology. In other words, birds are a barometer of everything that we do right in our gardens. If that is the case then we are doing some things very right at Longmeadow because it fairly teems with birdlife and watching them at work is one of the great bonuses of the winter garden.

The whole relationship between the garden and birds changes as soon as the leaves start to drop. For a start they become more visible. They crowd the branches as a series of shapes rather than sounds. The outline of a small tree will suddenly break as a flurry of birds leaves, scared away from grabbing berries whilst they can. It makes you aware of how present the soft midsummer sound of unseen birdsong is, how important an element in any garden. Winter bird sound is much harsher, a series of warnings rather than wooings. Occasionally a robin will astonish the afternoon with a burst of song, but a November afternoon in this garden tends to shuffle with staccato sound, like overhearing an argument in another room.

Winter here is heralded by the arrival of the fieldfares and redwings, just as surely as summer is certified by the first swallow. But whereas the swallows, supple as Mercury, arrive with a kind of soaring familiarity, the fieldfares are a curious mixture of awkward truculence and shyness, rising in a clucking, chattering cloud if you so much as appear within their sight and yet always pushing aggressively forward as soon as they think your back is turned. Everything about them is harsh and jerky, yet I like them. They are of the season. They like the apples left in the Orchard best of all and will fiercely defend a tree with windfalls from other birds. They also do a lot of good, as they eat snails, leatherjackets and caterpillars.

The other winter thrush, the redwing, is smaller, daintier and altogether less intrusive. Whereas the fieldfare has an instantly recognisable grey/mauve head, the redwing is only distinguishable from a song thrush in flight when the red flash under the wing is visible – although its tendency to flock, like the fieldfare, is a giveaway.

Birds belong to the garden as much as the plants. A healthy bird population is an important link in a healthy garden, not to say a healthier, happier garden.

I like to cut back summer's growth from the borders by degrees, leaving as much as possible for as long as possible.

DECEMBER

I have always considered December in the garden to be a kind of horticultural anti-matter. At best the garden waits patiently and at worst it sulks and collapses in on itself. I do not blame it at all – indeed I share all these feelings myself.

It is all about the light. The first three weeks of December take a bad situation and make it steadily worse. We are both counting the days till the 21 December when there is the comfort of knowing that there is no further to fall. The year has reached its nadir. Then Christmas comes with its lovely vulgarity and busy domesticity, with the ritual of collecting mistletoe from the Orchard and cutting holly from the more straggly specimens across the garden. Above all there is a profound sense of having come through with just enough resolve to start to pick up the pieces and start again.

Too gloomy? Of course, much too gloomy by half, but that, of course, is the problem, not the cause. Clear, clean cold can come to the rescue, never more dramatically so than in December 2010 when there were days that the whole garden (down to the last twig and seedhead) were bejewelled with ice and snow under the bluest skies of the year. This is a guilty pleasure because beneath that sparkling carapace plants are suffering and snow always leaves a horrible mess on our heavy soil, but it is a price that I am happy to pay for a while. In fact, any amount of cold (it got down to -18°C (0.4°F) on Christmas Day last year) is more or less welcome if it can come without snow because it means that the mud is locked solid, the ground is hard and everything is accessible until the thaw.

One of the reasons that we have planted so much evergreen structure throughout the garden is to combat the dreadful brown and grey gloom of midwinter. The hedges of box, yew and hornbeam and the spaces that they define and enclose are the very opposite of monotonous structure but vibrantly defiant, shoring up the garden as well as my own ruins.

The truth is that we do remarkably little gardening this month. On warmer days we collect fallen leaves, plant the last of the tulips, as well as the occasional tree or deciduous shrub, although there is little room for these here any more. But mainly it is housekeeping work such as fixing fences and tools, tidying the potting shed and washing pots. And just as the December garden seems to take a break, so do we.

CHRISTMAS TREE

It is the tree-planting season outside and the tree-dressing season indoors. The latter, of course, refers to Christmas trees, and Christmas would not be Christmas without a tree of some sort decked in whatever finery and baubles feel suitable. A fake Christmas tree is better than no Christmas tree, but I love the resiny smell of the real thing, the slightly scratchy feel of the needles (which of course are merely leaves that have evolved to cope with drought and cold) and that sense of wonder of having a tree indoors.

Until the middle of the nineteenth century the dominant evergreen trees were yew, holly, box and juniper. A few trees such as the Norway spruce and cedar of Lebanon were gradually being introduced but it was not until the nineteenth century that there was an explosion of new evergreen trees imported to this country, and this coincided with the introduction of dressing a Christmas tree – brought in from Germany by Prince Albert in the 1840s. There is no reason why any evergreen could not still serve as a Christmas tree but the field has narrowed to three main species that dominate. All of these are good and all, if purchased with healthy roots, can be planted outside in the garden once they have served their turn brightening up your Christmas.

The three species are Norway spruce (*Picea abies*), the Nordmann fir (*Abies nordmanniana*), and the Colorado spruce (*Picea pungens*). All three have specific virtues and can last for a long Christmas season if looked after properly. All three will also grow in most gardens if they are bought with healthy roots and planted carefully as soon as possible after Christmas.

The Norway spruce, *Picea abies*, the archetypal Christmas tree, has over 350 cultivars that are drooping, upright, dwarfing, golden, prostrate and in any other arboreal configuration you can imagine. It grows very fast and for centuries it was the main source of softwood, or deal. Although almost everyone nowadays only comes into contact with it as a tree small enough to fit into the living room, given the right conditions, it will grow to more than 60m (200ft) tall and is officially Europe's largest native tree. It is completely frost-hardy – even when very young – so ideal for cold, exposed sites. It prefers a slightly acidic soil and really does not like drought, salt spray, or growing on chalk or limestone.

The Nordmann fir, *Abies nordmanniana*, has rightly become extremely popular as a Christmas tree over the past decade or so. It tends to have denser and more horizontal branches than the Norway spruce and is less willing to shed its needles when put in the warm environment of a centrally heated house. This is because it is more truly evergreen than the Norway spruce in that it only sheds and replaces its needles after about 15 years. It was introduced to Britain about 150 years ago and will grow even bigger than the Norway spruce, reaching 68m (225ft). It grows on limestone in its Caucasian home but, rather curiously, like the Norway spruce, it grows best in moist, cool, slightly acidic conditions in this country. However if you do garden on chalk and wish to have a Christmas tree outside in the garden then this is a better bet than the Norway spruce.

I like the Colorado, or blue spruce, *Picea pungens*. Although it comes from the southern states of America, it originates from a high altitude, so is very hardy and grows into a tall, very straight, beautiful tree, with glaucous blue needles the colour of cardoon or artichoke leaves. The high altitude and bright mountain light give it a rather stiff habit – which is one of its main attractions as a Christmas tree in that it is much easier to decorate. The cultivar 'Kloster', bred just over 100 years ago in Holland, has, along with 'Moerheim', recently become a Christmas favourite and both are now grown just for use as Christmas trees. Both were bred to accentuate the blueness and stiffness of the parent plant.

Keeping your Christmas tree spruce

- The biggest problem with a Christmas tree is the shedding of its needles. It does this as an emergency response to being too dry and too hot. Even background central heating is uncomfortably hot for it. A draughty hallway is ideal. So the absolute rule is to keep it as cold as possible (it would be perfectly snug in a deep freeze) and water it.

- Get one with roots if possible, even if you are not intending to plant it. Then 'plant' it in a bucket or trug of sand and keep the sand watered. The tree will draw up the water via its roots.

- If you buy one without roots – and most of us will – buy a cast-iron tree holder to support it. These have a well for holding water. Keep this topped up throughout the festive season. Your tree will draw water up through the cut trunk just like a cutflower in a vase.

HOLLY

Everyone knows the Christma carol 'The holly and the ivy / when they are full grown / of all the trees in the wood / the holly bears the crown'. But when was the last time that you were in a wood and noticed the holly? For most of us gardeners, holly is something that we tend to clip and trim into a hedge or topiary and I suspect that even the average non-gardener has reduced the world of holly down to Christmas sprigs. But it is a wonderful and rather magical tree with a place rooted deep in British folklore.

There is a wood I often visit, tucked away on the Hereford and Monmouth border, where the holly trees are ancient, huge and extremely beautiful in a venerable, craggy, slightly awe-inspiring way. In some of them the wood of the original trunks has rotted and hollowed to a grey husk that a man could easily hide inside. But new growth sprouts from the old wood to make thick, strong branches. Some of the branches are astonishingly long and whippy and almost leafless save where they grow out into the light. Put all this together and you have a curious, ruined kind of tree capable of living like this to a huge age.

One of the reasons hollies on farms can look like this here on the Welsh Borders is that for centuries they have been harvested for fodder. Rather astonishingly, given how prickly most holly is, it makes highly nutritious and desirable winter feed for

Not all holly is prickly. This **Ilex × altaclerensis** *'Camelliifolia' growing in the Box Ball Yard is quite smooth. Although holly casts a very dry shade it does best in moistsoil and a sunny site.*

sheep and cattle when there is no grass and it saw many a farmer through periods when snow or ice covered the fields and their hay stocks – never very plentiful in this part of the world – were running low. In fact, many holly leaves are rich in caffeine and *Ilex paraguariensis* is brewed to make mate, a kind of Argentinian tea. 'Hollins' are stands of holly trees grown and lopped for centuries specifically for this purpose. It underlines the adaptability of the plant, its toughness and willingness to be hard-pruned.

Holly is ideal for topiary and hedges and like yew and box will re-sprout from the bare wood. This means that an old overgrown holly hedge can be 'halved' – by which you cut right back to the stem on one side, wait for signs of healthy regrowth and then a year or two later repeat the operation on the other side of the hedge. That way you dramatically reduce the width of the hedge whilst at the same time increasing its density.

Holly provides perhaps the deepest shade of all British trees and this can lead people to think that it will grow in shady, dry conditions but in fact it does best in a sunny site in rich soil.

Although there are over 400 species of holly there are easily the same number of cultivars and hybrids. But the basic holly that we think of as Christmas decoration is the English holly, *Ilex aquifolium*. The Latin name, *Ilex*, was originally used by the Romans for the holm oak, *Quercus ilex*, another evergreen tree whose leaves bear some resemblance to holly, and the word 'holm' is derived from holly, but there is no family connection. English holly is characteristically spiky, has red berries and can grow to a really substantial tree of 24m (80ft).

Some hollies are fiercer than others. Hedgehog holly, *Ilex aquifolium* 'Ferox', has prickles on its surface as well as round the edges. Other hollies have leaves on higher branches as smooth as a laurel, especially the Highclere hybrid, *Ilex* × *altaclerensis* 'Camelliifolia'. The male form is notable for its vigorous purple shoots and all forms tend to have larger, less spiky leaves than English holly and be more vigorous.

Most holly trees are either male or female and to get berries on holly you must have a mate in your garden to fertilise the flowers (although this need not be of the same variety). This task is made rather tricky by the names given to several of the most popular varieties. So 'Golden Queen' is actually male and 'Golden King' is female. If you do get it right – and you might be saved by the presence of a holly of the right sex in a neighbouring garden – then your berries may not always default to the colour red.

Ilex aquifolium 'Bacciflava' is a female tree with very bright yellow berries. I have had one in the Spring Garden and it happily produces its yellow flush of berries although it must be pollinated by one of the red-berried varieties also in the garden.

But at Christmas I think your holly berries really have to be red. The shiny green leaves and bloody berries throb with aggressive life and bringing them indoors is an act of defiance and faith that the shortest day – 21 December – will pass and be the door to spring and regrowth.

MISTLETOE

Mistletoe grows well in my garden – I have five good bunches of it – but its success has absolutely nothing to do with me. It is untamable and unmanageable, growing where and when it chooses. In fact, I have managed to persuade some to grow by squidging a berry into the fissure in the bark of an apple tree, but that was just one out of perhaps 20 such attempts and I have absolutely no idea why that particular one worked and the others failed or even if mistletoe would have appeared on that particular tree anyway, without my busy interventions.

Of course it does not just appear like magic. We know that it is spread by birds, mostly from the thrush family, that eat the berries, find the gooey white flesh that surrounds the seed sticking to their beaks which they then wipe clean on a branch. If there is seed in that residual goo and if the tree has bark with a suitable texture to catch and hold that seed and if the branch is of the right size and if the wood the right texture, then it has a chance of germinating and growing. But you can see that there are a lot of variables. It is also suggested that the plant is spread by the seed passing from the other end of the bird and being deposited on a suitable branch along with a dollop of fresh manure to help it on its way, but there are doubts about whether the seed could survive the acidity of the bird's digestive system.

The name derives from the Saxon *mistel*, which seems to derive both from the plant and dung and perhaps the mistle thrush, combined with *tan* meaning a twig. Either way the thrush continues to spread the seeds of mistletoe far more effectively than the most ardent gardener.

Mistletoe, *Viscum album*, is an epiphytic semi-parasite. That is to say that it lives on other trees and at least partly depends upon them for its source of nutrients. It has haustoria instead of roots, which grow into the host tree and tap into its xylem, the woody layer that lies beneath the phloem and bark and which transports water and

minerals from the roots out to the leaves. Mistletoe nearly always grows on smaller branches and will exert greater osmotic pressure than the tree itself so therefore can suck out the mineral-rich sap before the tree's own leaves can. As long as there is a good supply of water and minerals and as long as the mistletoe does not grow too large, this relationship can be perfectly amicable with the tree having plenty of goodness to spare.

But it always ends badly. As the mistletoe flourishes it needs more and more water and so deprives the tree and causes stress. The tree reacts to the presence of the haustoria by creating big knots – which is a problem if it is being grown for timber – and will react to the invasion by growing a mass of twiggy branches known as witches' brooms. Because of the mistletoe's greater transpiration than its host, cavities can form in the xylem tubes that carry the water up the tree causing water molecules to break apart – exactly like an airlock in a heating system. This stops water reaching the branch beyond the growing parasite, weakening and potentially killing it. Finally the success of the mistletoe is its own undoing because the haustoria become so dense that they block the supply to the branch beyond it which then dies back, and the mistletoe is therefore deprived of its source of nutrients and effectively kills itself.

In fact, mistletoe has evergreen leaves so can photosynthesise and therefore provide some of its own nourishment. This means that a tree can carry mistletoe for many years in a perfectly easy relationship, particularly if there is a good water supply. This is why you are much more likely to see mistletoe in the west of Britain – where there is much more rainfall – than in the dry east. Here, on the Welsh Borders, it is everywhere and seems to be spreading. In fact, the biggest source of British mistletoe for the Christmas market comes from one sale in Tenbury Wells, on the Shropshire/Herefordshire border, that is held at this time of year and to which all the local farmers bring great bundles of mistletoe to auction as a nice little Christmas earner.

It is picky about its host. Not any tree will do. Hawthorn is by far the favourite although it also grows well on apple trees, limes and poplars. The latter, being very fast-growing and often planted along roadsides, are the most visible source of it for the passing motorist and are often decked with great balls of mistletoe like giant Christmas trees. Despite the mythical Druidical connections, it is rarely found on

Mistletoe is more common along the Welsh Borders than anywhere else in Britain and is becoming established in half a dozen trees in the Orchard.

oak. It is also rarely found in woods and if it is, will always be in a glade or clearing. This is consistent with its liking for trees standing alone in fields, orchards, gardens and narrow plantations.

The berries start out the same woody olive colour as the stems and leaves before becoming the familiar milky, opaque bubbles stuck onto stubby nozzles of branch. They ripen erratically and at different times from year to year, from November to April. Not all mistletoe carries berries. It is dioecious, which means that the male plants are separate from the female ones, and only the latter have berries. This explains why a very healthy ball can be berryless whereas another nearby can be much smaller but festooned with them. For example, only two of my five bunches ever produce berries.

Mistletoe's connection with Christmas, like so many seasonal traditions, originates in our pagan past. Remember that until around 1800 – a mere blink away in the greater scheme of things – few evergreen trees grew in Britain. So a mysterious plant that had olive-green leaves in the dead of winter, that just appeared suspended between earth and air, and which had unusual milky white berries after almost all other fruit had gone, was likely to be revered. Perhaps the Druids held oak-borne mistletoe in such awe precisely because it is so very, very rare. In any event, a sprig of mistletoe protected the house from goblins and sorcery which were most threatening at the time of the winter solstice – 21 December. Its magic spread to medicine and the plant was variously used to cure jaundice, whooping cough, adenoids and epilepsy.

It was always considered very unlucky to decorate the house with mistletoe before Christmas Eve although in the west of England the bunch remains untouched until the new one is brought in to replace it. Kissing beneath the bunch is simply a remnant of fertility rites: it was babies that were intended rather than permission to grab a seasonal smooch, so think on that when you tipsily hold your kissing-sprig over a selected victim.

RED CABBAGE

I like to have as much from my garden as possible for Christmas dinner. I have yet to hand-rear a goose or turkey but we can and do gather potatoes, onions, sprouts and red cabbage from the garden. The latter is the most essential of the lot. Simmered very slowly with brown sugar, juniper, onion and vinegar it is the perfect accompaniment to turkey or any game and improves in the week from Christmas to New Year, being somewhere between a vegetable dish and a pickle.

But first grow your red cabbage. I like 'Red Drumhead' which has a deep, almost purple, colour and 'Red Dutch' is also good. Like all cabbages, they are slow to grow and mature, so to be at their best for Christmas should be sown in May and certainly no later than Midsummer Day.

I sow mine in seed trays and then prick the seedlings into plugs, growing them on in a coldframe before planting them out into their final positions. You can equally well sow them into the soil of a seed bed, but be sure to thin ruthlessly so that they are at least 8cm (3in) apart by the time they are 8cm (3in) tall. This will help establish healthy roots from the first.

Transplant them to their growing bed when they are about 15–23cm (6–9in) tall which will be some time in July. Space them 45cm (18in) apart in a grid and firm in well. Cabbages should follow on from a legume crop, like broad beans or peas, and will not need any extra manure. Keep all cabbages covered with a fine net to protect them from pigeons and, above all, cabbage white butterflies. They should be ready for harvest from November through to May.

CHRISTMAS HERBS

By Christmas time it can feel as though the garden has closed for the winter. But there is a part of my garden that is still working hard, albeit in a limited manner, and this is the Herb Garden. In fact, a few herbs are as vital to Christmas dinner as the vegetables I grow to accompany the turkey.

Sage and onion – and chestnuts – are the traditional stuffing for a turkey and are the best I think. Anyway, Christmas is a time for enjoying traditional rituals and recipes, not experimentation. Onions are long stored in a cool shed. Chestnuts (*Castanea sativa*) make a wonderful tree but need a soil more acidic than mine to grow and thrive. But sage (*Salvia officinalis*) should, and could, be grown in every garden to be picked fresh on Christmas morning. Sage is a Mediterranean herb and therefore likes its native conditions replicated as closely as possible in your garden – which is admittedly tricky in the middle of a gloomy northern December. The key to growing it successfully is sunshine and drainage. Like most Mediterranean herbs, it grows better in a poor soil. Too much goodness makes the foliage and young stems too soft and sappy whereas they have adapted to hot, dry conditions where hard woody stems and small leaves are an advantage. We grow a large-leaved sport of *Salvia officinalis*, purple sage (which is more tender and less good for cooking) and narrow leaf sage, *Salvia lavandulifolia*, which has, I think, the best flavour for cooking.

Thyme, *Thymus vulgaris*, is not part of turkey stuffing but goes very well with the bird and is excellent mixed with butter and slid under the skin to moisten the flesh as it cooks. Thyme also likes Mediterranean conditions but is even fussier about drainage and shade than sage. It really must not have any shade at all – even from itself – so I always grow some in a pot so it can be moved out of the shade of any neighbouring plants.

Other than the basic *Thymus vulgaris*, which I regard as essential for the kitchen, I strongly recommend lemon thyme which has a deliciously lemon fragrance and grows very well. To have good quality thyme at Christmas you must bring the pot indoors around October or cover the plants growing outside to protect them from cold and rain. But it is worth the trouble for that buttery, herby flavour with the turkey.

As is well known, turkey is a relatively recent Christmas dish for most people and the medieval Christmas dish of choice was a boar's head, although it might be tricky finding such a thing at your local butcher or supermarket. This was served traditionally with another Christmas herb, rosemary.

Rosemary was considered a symbol of the Nativity and used to decorate churches at Christmas. Rosemary is a tough shrub but has a nasty habit of suddenly turning brown and dying within months. I must have lost a dozen established plants like this over the years. The main cause seems to be the combination of midwinter cold and damp. It absolutely hates sitting in cold, wet soil and if you see a branch turning brown and dying back this is likely to be the reason. If you grow rosemary in a pot I recommend the very upright variety 'Miss Jessopp's Upright'.

For those who like to celebrate Christmas with a good joint of beef (and a rib of beef is better than any turkey) then the best accompaniment is fresh horseradish from the garden. Horseradish (*Armoracia rusticana*) has large and luxuriant leaves in summer but it is an herbaceous perennial so these completely disappear by mid-December, although the long taproots are still there beneath the soil. If you dig them up in autumn they are still relatively mild but by Christmas they will have achieved some real bite. Certainly homemade horseradish sauce is much, much better than anything I have ever bought.

It grows best in heavy soil (but it is easier to dig up in lighter soil) and it can become a weed if not restricted to an odd corner. We have ours growing in the Damp Garden and – very happily – in the stony driveway where we park our car!

New plants grow very easily from root cuttings. Dig some good straight roots and cut them into 5–8cm (2–3in) lengths. Pot them up until new growth appears and then plant outside. In truth, even this is unnecessary as it is hard to stop it growing in almost any circumstances. Stick a piece of root in the ground and you are assured a supply of horseradish for many Christmases to come.

CELERIAC

Christmas would not be Christmas if it did not include chestnut and celeriac soup on Boxing Day – a dish as fixed in the Don calendar as a tree with lights on it. We do not grow our own chestnuts – our soil is not really acidic enough for them to be comfortable here – but I do grow celeriac with real passion.

Celeriac is never going to be the star of the show. There is no dramatic revelation when you lift it as there is with new potatoes or carrots. The swollen base is warty with roots and can be, I admit, disappointing. Many flatter to deceive with a seemingly generous top of a good-sized ball lurking in the soil, which reveals itself to be the top of a very flat plate rather than the melon-sized monsters that shops sell. Or, when trimmed of its tentacles of root, it is whittled down to something like a squashed golf ball.

No matter. I always forgive it and only blame my inadequacies as a gardener. It demands little other than rich soil and a good supply of water, and reliably grows into a workable vegetable. I say workable because its realm is the kitchen, not the garden or show bench. With the stalks, foliage and roots all lopped off and scrubbed under a cold tap with a good bristly brush to get in all the crevices, it emerges like an overgrown truffle – and in my opinion just as wonderfully tasty.

Celeriac (*Apium graveolens* var. *rapaceum*) is a member of the same family as celery, parsley, carrots and parsnips and if you leave a plant to flower, which it will do in the

spring after sowing, you will see that it develops an umbellifer flowerhead like cow parsley or dill. Whereas celery has been cultivated for medicinal purposes for thousands of years and first leaves and then the stalks have been eaten as a vegetable since medieval times, celeriac did not reach this country until the late 1720s and has never really entered the popular culinary imagination. This is a shame because it is delicious and nutritious, unlike celery which, when tough, probably needs more energy to digest than it supplies.

The leaves are edible and the French, who know a thing or two about cooking, carefully keep them and use them to add a celery flavour to soups and stews. We tend to discard them, hopefully in the direction of the compost heap rather than the landfill site.

You have to give celeriac your time and attention although this is hardly demanding. It needs peeling – but the peel can be dried and kept to flavour stocks and soups – chopping and cooking. It also rapidly oxidises so must be put into a bowl of water with a squeeze of lemon juice once prepared. But if you can bear the extreme rigours of dealing with the thing itself rather than some pre-prepared travesty then, as well as being delicious roasted or puréed with lots of cream – delicious cold, eaten straight from the fridge – celeriac adds an earthy, musky taste to soups, stews and mashed potato.

In the same way that celeriac is truculently reluctant to be fast in the kitchen, it takes its time in the garden. The swollen base does not develop until the leaves are well established and needs good quality soil and plenty of water. Soil with lots of organic matter will make a huge difference both in retaining water and in helping it to develop a good root system that will grow deep to find all available moisture, which can dramatically reduce the watering you need to do.

I sow the seeds in March, which, like celery, are tiny, so I scatter them thinly on a seed tray (do not cover them) and then prick them out into plugs. These then get potted on into 8cm (3in) pots. Although they are quite hardy, it is a mistake to put them out too early because they will not grow in cold soil and whilst they sit, waiting like the rest of us for spring warmth, the slugs tuck into them with relish. So I wait until mid-May before planting them out a generous 30cm (1ft) apart in blocks or

Celeriac will never win a beauty contest but it is a subtle, delicious and adaptable vegetable and I always grow lots of it.

rows. I weed and water regularly until late summer when I start to remove any foliage that is not dead upright. There is no science to this but it helps get as much rainfall directly to the roots rather than settling on and evaporating from the foliage.

Last winter my celeriac froze solid and I harvested them with a pickaxe, but they were usable when thawed. They do need some protection from extreme cold, however, and I vacillate between mulching them well with straw or bracken (which is good in a mild winter but encourages slugs) and lifting them and bringing them into a frost-free shed (although they dry out easily and need storing in leaf mould or moistened vermiculite) in a similar way to dahlias.

BRUSSELS SPROUTS

Although there seem to be references from Brussels in the thirteenth century and mention of them in Burgundian feasts of the fifteenth century, Brussels sprouts have only become commonly grown in English and French gardens since the seventeenth century. The first documented British recipe is from Eliza Acton in 1845, and it seems that sprouts only became popular after World War One. So the 'tradition' of serving them at Christmas with the turkey is, like most of our cherished traditions and institutions, a relatively modern invention.

The key to reliable crops is a long growing season and firm soil. I sow my seeds into seed trays in March, pricking them out as soon as the first 'true' leaves appear. They need a decent depth of soil – at least 8cm (3in) – so that the seedling can develop a strong taproot. This will come into play when the young plant is transferred to the outdoor growing position in May. Leave at least 60cm (2ft) between each plant and if you have the space 90cm (3ft) is better. They can be sown into an outdoor seed bed in April, but they must be thinned early so that there is at least 8cm (3in) between each plant.

As with all brassicas, sprouts grow best on alkaline soil that is well cultivated but not freshly manured. Following on from French beans cleared in the previous autumn is ideal or, slightly later on, in early summer, use ground vacated by an early crop of broad beans or peas. Fork the ground over, removing any weeds, and then tread it firm as though preparing for a lawn. This might seem counter-intuitive but it encourages the plant to root much more firmly. I earth up each plant with a hoe in midsummer to encourage more roots to grow and anchor them firmly to the ground, but I usually end up staking each plant with a cane as well. The point being that a healthy Brussels sprout plant is inevitably top heavy and will topple or lean.

As autumn progresses, remove leaves that start to yellow or if the plants are becoming crowded. Use your common sense over this but in general the colder and more exposed the site, the more leaves you should leave on as they do provide some insulation for the sprouts forming in their lee. Once the sprouts start to form and are the size of marbles you can cut off the top of the plant to make them all ripen at the same time. This is only an advantage if you wish to pick them as one commercial harvest or to freeze. Conversely the top can be left as a perfectly tasty cabbage in its own right to be harvested later in the season.

Cold improves the flavour down to about -10°C (14°F), so the best pickings are likely to be after the first hard frosts. But if the weather looks like being very fierce they can be harvested and stored by cutting the whole stem, removing the leaves, and hanging in a cool, dark place for up to a month.

Red sprouts are good-looking enough to grow for purely decorative reasons, and they always feature in the Ornamental Vegetable Garden but they do not taste as good as most green sprouts.

Taking stock

The days are still desperately short, the weather miserable and the party over, but in many ways Boxing Day is one of my favourite days of the year. The pressure is off and I have a profound sense that things will steadily get better. I know full well that there will be all kinds of blips and bumps on the way but the road to spring and all its joys starts right here.

The first thing I do in preparation for the coming gardening year is to drink a large dose of honesty. This can be a harsh medicine but the benefits are fast-working and long-lasting. I go out and take stock. I own up to all the errors and omissions of the past year and I look the garden straight in its midwinter eye. I try not to gloss over the mud, the sprawling weeds or the uncut hedges. I recognise the unpruned tangled shrubs and fruit bushes, the dozens of pots standing out with unloved plants and the trays of seedlings that never got pricked out in time. Then I start to tackle them, bit by bit. I do not do all of this on Boxing Day of course. Heaven forfend. Mooching about, taking stock, taking my time is quite enough. And it is not all self-flagellation. I acknowledge the good too, trying to work out what is working and what I need to value most.

Out of this – and it spreads over a few days between Christmas and New Year – a plan emerges which I will base most of the coming year's work on. This is not just a bunch of good intentions but a real plan with lists and dates. More organised gardeners than me would have probably placed all their seed orders by now but from this scheme I start working out what I need to order (I still love the magic of seeds coming through the post and bless the internet for the way that this is all easier than ever) and so I work out a timetable of weeding, digging, mulching, sowing and planting. If this sounds unfeasibly organised then I am telling it wrong. I am a bumbler and fumbler and only get a lot done by doggedly plodding rather than any kind of whizz-kid fast tracking. But however small your garden or ambitions, a plan is a good idea at this stage of the year.

Presents

The truth is that most gardeners have the basic kit and are pretty particular about what they do and do not like. In short, giving – and receiving – presents for the garden is a minefield. As a long-time receiver of presents destined for the charity shop rather than the potting shed, here are some guidelines. Start by applying the hand baggage rule. It works like this: anything considered too sharp or dangerous to take with you as hand luggage on an aeroplane is likely to be desirable.

There is nothing remotely sinister about this. It is exactly the same enjoyment that you have from a good kitchen knife or a hand chisel. Secateurs, garden knives, loppers and scythes, even spades and hoes, are all as good as the edge that they will hold. If the steel is good it will stay good for a long time and easily be whetted by a few skillful wipes of a sharpening steel or stone. If that hand-forged blade, as sharp as a razor and fashioned exactly to the shape that best suits the job you are using it for, also is of a shape and weight that feels just right in the hand, then you have something of real beauty and mysterious power.

Ideally, one would only use hand tools that had this magic so that every action, every use, would be empowered by it whether it was cutting a length of twine with your pocketknife (and almost nothing is more useful in the garden than a sharp knife that lives in your pocket) or scything a meadow. The relationship between a careless mastery of a tool and the perfect implement for the job is one of the enduring pleasures of the garden.

You can never go wrong with a good pair of secateurs. You should always have a pair with you whenever you are in the garden. I have tried many over the years and currently use a Japanese pair, but you tend to get what you pay for. Holsters look a bit silly and pretentious but the alternative is a hole either in your hand or your pocket or both. Holsters make a good present in their own right.

Any true gardener needs twine. You use it all the time and Christmas comes round just in time to replace your stocks. Soft green twine is best for tying up all soft climbers, be they clematis or tomato, and tarred twine perfect for things like trained fruit trees or binding canes together. The point of twine is that, unlike plastic or wire ties, it does not cut into the delicate growth that it is helping to hold up. It also biodegrades, so can be added to the compost heap.

Do not turn your nose up at giving a few packets of seed even to the expert gardener. Some gardeners can be snooty about this but more fool them. You cannot go far wrong with easy vegetables like rocket, radish or carrot and if you know that someone is a keen veg grower try giving something a bit unusual like purple-podded peas, black radish or red Brussels sprouts. Give herb seeds like chives or parsley to those people who have a window box or even a kitchen windowsill and for those with a larger garden you can never have too many annuals like opium poppies, sweet peas, marigolds, cosmos or tobacco plants.

High quality seed trays are always useful. The test of quality in these things is rigidity. The test is to pick them up at a corner with one hand. If it bends or twists even the slightest then it is no good.

A garden diary is both useful and can be very attractive. I tend to use a plain office day-per-page diary to record what I planted and sowed, what flowered, what was harvested, what the weather was up to and any other observations that seem important.

Of course, the present that I shall be giving to myself is the most valuable of all and that is time. A few blissful days in the garden between Christmas and New Year will make me happier than anything else wrapped up under the tree.

A good knife is as important a part of any gardener's kit as a fork or spade. So is twine – although this lovely hop twine is becoming increasingly hard to buy.

LONGMEADOW

KEY

1 WALLED GARDEN
2 DRY GARDEN
3 COURTYARD
4 BOX BALL YARD
5 HERB GARDEN
6 LIME WALK
7 SPRING GARDEN
8 ORNAMENTAL VEGETABLE GARDEN
9 DAMP GARDEN
10 LONG WALK
11 JEWEL GARDEN
12 GRASS BORDERS
13 MOUND
14 COPPICE
15 COPPICE
16 COPPICE
17 WRITING GARDEN
18 CRICKET PITCH

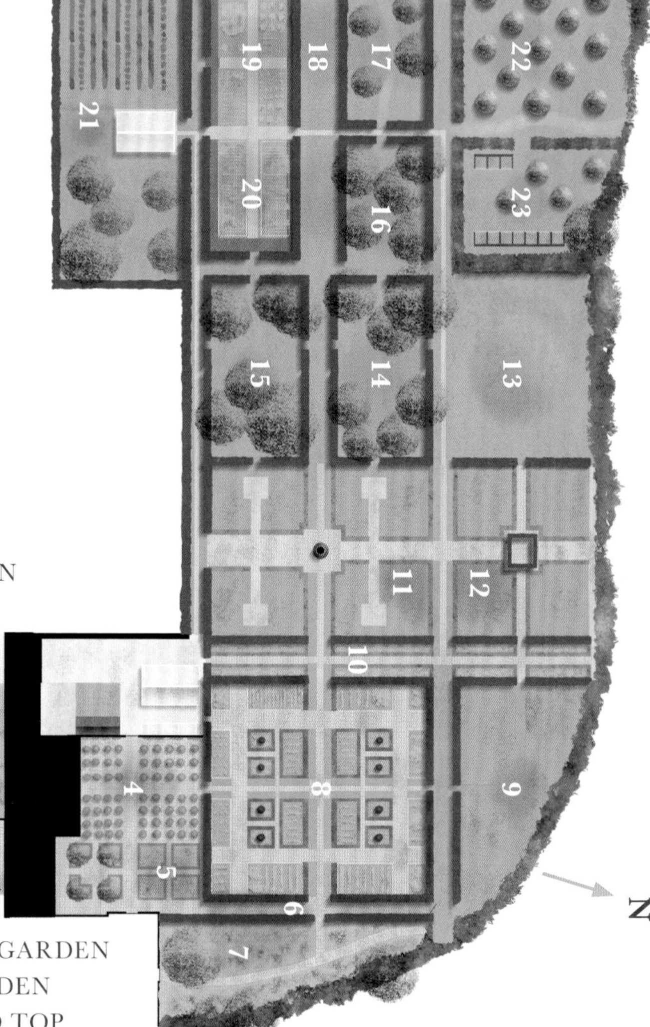

19 TOP VEGETABLE GARDEN
20 SOFT FRUIT GARDEN
21 STOCK BEDS AND TOP GREENHOUSE
22 ORCHARD
23 COMPOST/ORCHARD

INDEX

Note: page numbers in *italics* refer to captions, page numbers in **bold** refer to information contained in tables.

A

acer (*Acer*) 146
achillea (*Achillea*) 279, 283, 321
aconite 13, 14, **333**
'Guinea Gold' 14
agapanthus *52*
alfalfa 106
allium (*Allium*) *10*, 94, 129, 130–2, *132*, 172, 262, 283, 329–30, **333**
anemone (*Anemone*) 83, 86, *86*, **333**
angelica (*Angelica*) 155, 172
A. gigas *266*, 268
annuals 25, 48–50, 150, 172, 181–2, 221, 237–9
Anthriscus sylvestris 'Ravenswing' 153
antirrhinum 48
aphids 65, 153, 173–5, 281, 291, 327
lettuce root 105
apple (*Malus*) 26, 80, 83, 315, 345–6, *348*, 368–76, *371*, 427
'Worcester Pearman' 348, 369, 371, 372
apple bitter pit 375
apricot 26
aquilegia (granny's bonnet) 133
artemisia (*Artemisia*) 283, 286
artichoke 203
arum **333**
ash 83
asparagus 162–3, 365
asparagus beetle 163
asparagus fly 163
asparagus rust 163
aster (*Aster*) 218, 265, 286, 313–14, 328–9, 385
aubergine 64, 290, 334–5, 361

B

balling 187
basil 24, *168*, 170, 172
bay 170, 406
beans 64, 218, 265, 300–2, 315
see also specific beans
bear's breeches (*Acanthus mollis*) 283, *285*
bedding plants 48–9
beech 389
bees 16, 31, 86, 216–18, 268, 278, 362
beetles 136, 163, 234
beetroot 65, 163–5, *164*, 244–6, *244*, 358
berries 361–2
see also specific berries
biennials 48, 64, 135–8, 206, 244, 289, 366, 392, 407
bindweed 40, 125, 126
biological controls 73, 153, 175
birch 389
birds 73, 83, 87, 173, 254, 343, 415–17, 425
black sooty mould 173
black spot 187, 225, 385
blackberry (*Rubus fruticosus*) 218, 350, 361
blackfly 109
blackthorn (*Prunus spinosa*) 80–2, 298, 350
blanching 338
bleeding heart (*Dicentra*) 93, *182*, 184
blight 159, 251, 327, 376, 385
blossom 80–3, *80*, 286
blossom end rot 159
bluebell 83, *86*, 87, 260
bog gardens 219
bonfires 414
Botrytis tulipae 98
box 392, 398, 405–6, 419–20, 424
Box Ball Yard *223*, 327, *422*
box (*Buxus sempervirens*) 8, 83, 310, 324–7, *327*, 353, *354*
box blight 327
bramble 270
brassicas 19, 40, 65, 66, 106, 136, 173, 208, 265, 281, 289, 315, 368, 407, 411
see also specific crops
broad bean 62, 65, 109, *109*, 173, 265, 300, 392, 429
broccoli 38–40, *38*, 65, 208
purple sprouting 38, *38*
Brussels sprout 65, 429, 434–5, 437
bud drop 296
buddleja (*Buddleja*) 30, 45, 270–1
bulbs 24
container-grown 398
naturalising 30
spring 53–4, 329–33, *333*
summer 94–6, 221
bullace (*Prunus domestica*) 350
bumblebees 16, 31, 86
burdock 201
bush bean 300
busy Lizzy 49
buttercup 124, 125, 126, 262
butterflies 38, 83, 136, 177, 193, *244*, 268, 270, 278, 281, 411–13, 429

C

cabbage 65, 136, 208, 281, 407, 410–13, *413*
red 429
calabrese 38
Calamagrostis x *acutiflora* 321
calendula 239
callicarpa 362
camellia 24, 173, 406
canker (*Nectria galligena*) 346, 373–5
canna 265–6, 278–9, 313, 385
cape hyacinth (galtonia) 95, 237, 286
cardiocrinum (*Cardiocrinum*) 95
cardoon 286

Carex 281, 321
C. comans 361, 398
C. elata 'Aurea' 321
carrot 62, 65, 153, 175, 201, 206, *287*, 289, 341–3, 358, 437
carrot fly 289, 384
caterpillars 281, 411–13, 417
cauliflower 65
cavalo nero 19, *19*
celandine, lesser 226
celeriac (*Apium graveolens* var. *rapaceum*) 65, 431–4, *433*
celery 64–5, 175, 206, 289, 315, 382–4, *382*, 431–3
celery leaf miner 384
cerinthe 283
chafer grubs 177
chard 36, 63–4, 358
cherry (*Prunus*) 26, 80, 82, 83
chervil (*Anthriscus cerefolium*) 155
chestnut (*Castanea sativa*) 429, 431
chickens 73, 369, 415
chickweed 70, 125, 127
chicory 165–7, *167*, 335–8, *335*, 407
chilli 380–2
chive 170, 172, 437
choisya 406
Christmas trees 420–2
Cistus 'Thrive' 283
citrus 338–41, *341*, 398
clematis (*Clematis*) 45, 191, 371, 437
early-flowering 43, 87–90, *89*
late-flowering 30, 43, 229–31, *231*, 265, 315
mid-season 43
planting 89–90
pruning 43
clematis wilt 231
climate change 77, 219, 265, 282
climbing beans *298*, 300, *300*
cloches 167, 247, 391
clover 65, 106, 176, 218
red 262
clubroot 65, 136, 368, 411
colchicum (*Colchicum*) 234, 322–4
cold conditions 24–5, 49, 385–7, *391*, 396, 398
coldframes 63
Colorado spruce (*Picea pungens*) 420, 421
comfrey (*Symphytum officinale*) 226, 257–59, *258*
compost 45, 71–3, *73*, 119–24, *119*, *122*, 258, 387–8
Coniothyrium hellebori 34
conkers 363
container gardening *197*
aubergines 334
citrus 338–40, *341*
and fertiliser use 71
herbs 172
pansies 60
pelargonium *197*, 198, 200
snowdrops 17
tomatoes 304
and winter 396–8, *396*
Coppice 9, 29, 30, 56, 58, *58*, *78*, 82, 83–7, *86*, 89, 90, 96, *109*, 155, 226
coppicing 83, 144, 146
cordons 26, 66, 371, 372
coriander 170, 172
cornflower (*Centaurea*) 49, 218, 261
cornus 86
Corydalis 184
C. flexuosa 93
Cosmos 150, *150–3*, 278–9, 286, 313, 385, 438
cotoneaster (*Cotoneaster*) 362
couch grass 40, 125, 126
courgette 64, 246–7, *247*, 378
cow parsley 129, 153, 175, 260, 262, 433
cowslip (*Primula veris*) 56–8, *58*, 260, 262
crab apple (*Malus*) 80, 82, 321
Crambe cordifolia 286
crane fly 176
Cricket Pitch 53, 138
crocosmia (*Crocosmia*) 95, 239, 266–8, *266*, 279
crocus (*Crocus*) 30, 53, 228, 332, **333**, 398
crop rotation 64–6, 106, 289
cucumber 64, 290–1, *291*, 334, 378
cucumber mosaic virus 291
cucurbits 247, 378
see also specific cucurbits
currant (*Ribes*) 66
black 286–7
red 256–7, 286, 287
white 286
cuttings 200, 391–2
black currant 287
box 324–5, *327*, 353, *354*
dahlia 274–5
hardwood 287
horseradish 431
lavender 197
oriental poppy 143
root 143, 431
sage 400
semi-hardwood 197
wallflower 366
Cyclamen hederifolium 322
Cylindrocaldium buxicola 327

D

daffodil (*Narcissus*) 47, 53, *78*, 228, 262, 329–30, *330*, 332, **333**, 368, 398
dahlia (*Dahlia*) 29, 239, *240*, 265, 271–5, *271*, 278–9, 313, 318, 385, 391, *392*, 434
daisy 70, 124, 261, 262, 278
Damp Garden 9, 95, 96, 124, 219, 226, *228*, *231*, 232, 235, 277, 316, 361, 376, *389*, 431
damson 80, 82, 285–6, 296, 298, 350
dandelion 70, 124, 125, 218, 262
daylily (*Hemerocallis*) 234
dead-heading 187, 271, 275, 391, *392*
Deschampsia cespitosa 'Golden Dew' 360–1
digging 40–2, 391

dill 153, 170, 172, 175, 433
division 16, 51, *52*, 94, 143, 328–9
dock 125
drainage 330, 338–9, 385
Dry Garden 9, 53, 140, 188, 242, 282–5, *296*, 360, 361
Dutch Elm disease 389
dwarf beans 112, 300

E

earwigs 275
Echinacea 281
Eckford, Henry 240
ecosystems 74–5
endive 165, 167, 338
English elm 389
eremurus (foxtail lily) 188
eryngium (*Eryngium*) 286
erythronium 234, 332, **333**
espaliers 26, 30, 346, 372
eucomis (*Eucomis*) 95–6, 237
euonymus 406
euphorbia (*Euphorbia*) 90–1, 98
evening primrose (*Acanthus spinosus*) 218, 235, 283
evergreens 310, 389, 398, 405–6, 419–22
see also specific evergreens

F

fairy rings 176
false oxlip 58
fat hen 125
fennel (*Foeniculum*) 153, 155, 170, 175, 283, 358
bronze 279–81
fenugreek 106
ferns 322, 398
fertiliser 71, 94, 158, 177, 257–60
Festuca glauca 398
fieldfare 415–17
fig (*Ficus*) 283, 341–3
firethorn (*Pyracantha*) 362
flame guns 127
flea beetle 136
flooding 219, 226, 282
Florence fennel (*Finnocchio*) 265, 295, *295*
Flower Garden 313
foliage
golden 321
purple 144–6, *144*
silver 286
foraging 350–1
forget-me-not (*Myosotis*) 98, 136, 392
foxglove (*Digitalis purpurea*) 48, 136, *136*, 138, 392
freesia **333**
French bean 65, 112, 265, 434
French tarragon 170
fritillary 53–4, *54*, *57*, 234, 262, 321, 330, 332, **333**, 398
frosts 24–5, 49, 385, 387, 396, 398
fruit cages 24
fruit trees 25–6, 30, 218
fumitory 93, 184
fungal diseases 34, 65, 98, 105, 159, 162–3, 167–8, 187–8, 225, 231, 234–5, 250, 274, 291, 327–8, 332, 346, 363, 373–5, 385
fusarium wilt 162–3

G

Gardeners' World (TV programme) 10, 113
garlic 62, 65, 170, 172, 221, *247*, 250, 265
geranium (*Geranium*) 93, 132, 262
G. 'Ann Folkard' *191*, 228
hardy (cranesbills) 188–91, 191, 262
giant cotton thistle (*Onopordum acanthium*) 315, *315*
giant fennel (*Ferula communis*) 153–5
giant oat grass (*Stipa gigantea*) 283, 321, *321*, 360
gladioli 279
golden hop (*Humulus lupulus* 'Aureus') *240*, 321, 365, *365*
gooseberry 66, *66*, 167–8, 286
goosegrass 70, 127
Grass Borders 153, 155, 201, 268, 277, 315, 358, 362, 396
grasses 283, 321, 358–61, *358*, 398
see also specific grasses
green manure 65, 106
ground elder 40, 124, 125, 126, 127
groundsel 70, 127
Guignardia aesculi 363

H

half-hardy plants 24, 49
hardening off 20, *21*, 50, 163–5, 242, 246–7, 274–5, 411
hardy plants 24, 49
Hart's tongue fern (*Asplenium scolopendrium*) 322, 398
haustoria 425, 427
haws 362
hawthorn (*Crataegus*) 7, 14, 58, 89, 129, 362, 388, 392, 427
hazel (*Corylus*) 7, 56, *78*, 83, 86, 108, *108*, *109*, *132*, 229, *231*, 388
nuts 343
purple 144–6, *146*
hebe 406
hedges 405, 422
box 8, 324, 325, 392, 405–6, 419–20
cutting 308–10, *308*, 325
deciduous 308–10
evergreen 310
holly 424
hornbeam 8, 388, 392, 419–20
yew 8, 9, 351, 419–20
helenium 239, 277–9, *277*, 313–14, 321, 385
helianthus 51
hellebore (*Helleborus*) 13, 29, 30–4, *33*, *40*, 392, 405–6
hemerocallis (*Hemerocallis*) 51
hemlock, lesser 226
Herb Border 98

Herb Garden *22*, 429–30
herbaceous perennials 51, 92–4
herbs 168–72, *168*, 429–31
Herefordshire Marches 8
heuchera 93
hippeastrum (*Amaryllis*) 330
hoeing 127
hogweed, giant 153, 155
holly (*Ilex*) 83, 310, 321, 398, 406, 419–20, 422–5, *422*
hollyhock 218
holm oak (*Quercus ilex*) 406, 424
honesty (*Lunaria annua*) 135
honey bees 86
honeydew 173, 175
honeysuckle (*Lonicera*) 13, 34, *34*, 193–4, *193*, 239, 260
hop (*Humulus*) *240*, 321, 365, *365*
hornbeam 8, 310, 388–9, 392, 419–20
horse chestnut (*Aesculus hippocastanum*) 138, 363
horse chestnut bleeding canker 363
horseradish (*Armoracia rusticana*) 430–1
horsetail 125
hosta 51, 96, 226, 235, 286
hot borders 266–8
hoverfly 153, 175, 268
humus 121
hyacinth 95, 237, 286, 332, **333**, 398

I

Indian balsam 226
insecticides 175
inula 278–9
iris (*Iris*) 129, *138*, 140–1, 283, 330, 332, **333**
ivy 218, 406, 422

J

Japanese knotweed 124, 270
Japanese shield fern (*Dryopteris erythrosora*) 398
jasmine 193
Jewel Garden 9, 53, 87, 90–1, 98, 124, 129–30, *132*, 133, *138*, 141, 143–6, 150, 181, 191, *191*, 229, *231*, 232, 235, *237*, 239–40, 242, *244*, 265, 275, 277, 279, 281, 285, 313, 317, *322*, 325, 358, 360, 365, *365*, 396
juniper 406, 422

K

kale 65, 208
kidney bean 300
Kiftsgate 144
kniphofia (red hot poker) (*Kniphofia*) 234, 268
kohl rabi 65, 407

L

lacewing 153
ladybird 153, 175, *175*
lamb's ears (*Stachys byzantina*) 283, 286
lamb's lettuce (corn salad) (*Valerianella locusta*) 63, 105, 106, 410
land cress 63, 65, 105–6
landscape fabric 126
lavender 172, 194–7, *197*, 283
lawns 68–70, *68*, 176–7, 261–2, 355, 392
leaders 25
leaf blister 296
leaf mould 94, 387–8
leatherjacket 176, 417
leek 19–20, 65
leek rust (*Puccinia allii*) 20
legumes 19, 64–6, 106, 208, 218, 300, 429
lemon (*Citrus*) 339
Lenten rose (*Helleborus* x *hybridus*) 29–31, *33*
leonotis (*Leonotis leonurus*) 48, 150, 279, 316
lettuce 40, 63, 64, 100–5, *100*, *102*, 306, 315, 357–8, 378
lettuce grey mould 105
Leylandii 310
light 13, 14
ligularia (*Ligularia*) *231*, 232–4, 235
lily (*Lilium*) 94–5, 193, 234–7, *237*, 330, **333**
  L. henryi (Turk's-cap lily) 316
  L. martagon 234, 235, *237*
lily beetle 234
lily disease 234–5
lily of the valley 234
lime (citrus) 339
lime tree 30, 388, 427
Lime Walk 8
Long Walk 132, 135, *366*, 405
loppers 45
lovage 172, 175
lupin 65, 286
Lyme grass (*Leymus arenarius*) 361
Lysimachia 321
  L. ciliata 'Firecracker' 124, *144*, 146

M

mahonia 13, 401–3, *403*, 406
mallow (*Malva*) 218
mange tout 108
manure 65, 71, 106
Marasmius oreades 176
marigold (*Calendula*) 321, 438
marrow 246, 378
meadows 30, 261–2
meadowsweet (*Filipendula ulmaria*) 260, 262
medlar 80
Melianthus major 286
melon 290
mibuna 63, 105, 306
Michaelmas 328
Michaelmas daisy *see* aster
Midsummer's Day (summer solstice) 179

mildew 167
American 168
downy 105
powdery 136, 188, 225, 247, 291, 328
millet (*Panicum miliaceum* Violaceum') 361
mint 170, 172
Miscanthus 321, 360, *398*
M. sinensis 'Zebrinus' 400–1
mistletoe (*Viscum album*) 419, 425–8, *427*
mizuna 63, 65, 105, 306, 358, 407
moles 176
monocarpic plants 153, 155, 268
moor grass (*Molinia caerulea* subsp. *arundinacea* 'Windspeil') 361
moss 70, 355
Mound, The 138
moving plants 391–2
mulching 45, 70–3, *73*, 124, 126–7, 257, 258, 391
mullein 283, 392
muscari 332, **333**, 398
mushroom 350–1
mustard 105–6
mycelium 176
mycorrhizal fungi 185

N

Nasturtium (*Tropaeolum*) 279, 281, 283
natural gardening 260–1, *262*
nematodes 73, 87
nerine **333**
nettle 125, 177, 226
nigella 49
nitrogen 71
nitrogen-fixation 64–5, 106, 208
Nordmann fir (*Abies nordmanniana*) 420–1
Norway spruce (*Picea abies*) 420–1

O

oak (*Quercus*) 389, 406, 424, 427, 428
okra 64
olive 387
onion 20–2, *21*, 62, 65, 265, 429
Onopordum 286, 315, *315*, 392
orach 144, *322*
orange (*Citrus*) 338
Orchard 9, *80*, *330*, 369–71, 373, *375*, 415, 419
oregano (marjoram) 170
organic farmers 106
Ornamental Vegetable Garden 8–9, *22*, 24, 36, *36*, 82, 162, 203, *208*, 213, *223*, 250, 289, 315, 345–6, 357, 435
overwintering 24–5, 385–7, *391*
oxalic acid 36
oxeye daisy 261

P

palm 387
pansy 58–60
parsley 170, *170*, 172, 175, 205–6, 289, 431–3, 437
parsnip 62, 65, 175, 206, 289, 358, 408–10, *408*, 431–3
paths 8–9, 126, 127
pea 64–5, 108, 162, 206–8, *208*, 218, 265, 429, 437
peach 26
pear 26, 30, 80, *80*, 82–3, 286, 315, 343–6, *345*, 373, 376, 377
pelargonium (*Pelargonium*) 197–200, *197*, *198*
P. 'Sarah Don' 198, *198*
penstemon 24, 278–9, 392
pepper 64, *380*
perennials 40, 70, 124–7, 385, 391
pesticides 216
petunia 49
Phasmarhabditis hermaphrodita 73
pheasant's tail grass (*Stipa arundinacea*) 281, 360
Phoma clematidina 231
phosphorus 71
phytophthora 338–9
pigeon 38
pineapple sage (*Salvia elegans*) 318, 400
pinto bean 300
plant supports 108, *108*, 109, *109*, 158
plantain 70, 262
plum 26, 80–2, 184, 285–6, 296–8, *296*, 350
poinsettia 400
pollen 218
pollination 346
poplar 389, 427
poppy 48, 49, 93, 261, 285, 286
annual 181–2
Californian 279, 321
opium *181*, 182, *240*, 437–8
oriental 129, 141–3
Portuguese laurel 406
potassium 71, 257
potato 65, 113–15, *115*, *116*, 162, 251–2, 358, 429, 433
chitting 29, 113
earlies 113, 251–2, *252*
maincrop 113, 115, *115*, 251–2
new 221, 251–2, *252*
potato blight 159, 251
primrose (*Primula vulgaris*) 29, 47, 57–8, *78*, 83, 86, 96, 226, 262
privet 310
pruning
coppicing 83, 144, 146
hedges 308–10, *308*
maintenance 187–8
pear 346
plum 296
raspberry 347
root 311
rose 187, 223
spring 42–5
topiary 351, *351*
winter 25–6, 385
wound painting 45

Psylla buxi 327
pulmonaria (*Pulmonaria*) 91–3
pumpkin 64, 246, 377–8, *378*
purple elder (*Sambucus nigra* f. *porphyrophylla* 'Guincho Purple') 146
purple orach (*Atriplex hortensis* var. *rubra*) 144
purple spot 163
pyracantha 406

Q

quaking grass (*Briza maxima*) 361, 398
quince (*Cydonia oblonga*) 80, 82, 346, 376–7

R

radicchio 165, 167, 407–8
radish 40, 65, 116, *116*, 378, 407, 410, 437
raspberry 254–6, *254*, 347, 350, 361
red spider mite 60, 291
red thread disease 176
redwing 415, 417
rhizobium 65
rhizomes 14, 86, 140
rhododendron 406
rhubarb (*Rheum*) 36, *36*
rocket 62, 63, 105, *116*, 118, 306–8, 357–8, 437
root cuttings 143, 431
root pruning 311
root vegetables 65, 66, 106
Rosaceae 80
rose (*Rosa*) 30, 80, 86–7, 89, 96, 132, 173, 221, 235, 239, 283, 286, 361–2, 371
  alba 184–5, 286
  bourbon 184
  centifolia 184–5, *185*
  classic 184–8, *187*
  climbing 44, 222–5, 392
  damask 184
  dog (*R. canina*) 80, 149, 362
  gallica 184–5
  hybrid tea 44
  mulching 45
  pruning 44, 187, 223
  R. cantabrigiensis 148
  rambling 44, 222–3, *223*
  rugosa 185, 286
  shrub 44
  species 148–9
rose hips 362
rosemary 170, 172, 219, 283, 430
Rubus cockburnianus 286
rudbeckia 265, 278, 278–9, 385
runner bean 65, 112, 300, 302, 358
runners (strawberry) 213–14, *215*
rust 163, 225

S

sage (*Salvia*) 24, 170, 172, 218, 317–18, *318*, *391*, 392, 400, 429–30
salad crops 19, 63, 306–8
  see also specific crops
salsify 65
sawfly 167, 168
scab 373–5
scabious (*Scabiosa*) 218, 262
scale insect 339–41
scilla **333**
sclerotia (*Sclerotium cepivorum*) 250
scorzonera 65
sea holly 201–3
secateurs 45, 437
sedum 283, 361
seeds 437–8
  annuals 48–50
  aubergine 334
  beetroot 163–5
  broad bean 109, *109*, 392
  carrot 289
  celery 384
  chard 64
  chillies 380
  collecting 391
  hardy 385
  honesty 135
  lamb's lettuce 106
  lavender 197
  leek 19–20
  lettuce 104
  onion 20, 22
  peas 108
  potato 113
  rocket 118
  salad leaves 306
  snowdrop 14, 51
  sowing out 62–3
  sowing under cover 63
  tomato 304
  when to sow 62–3
sets, onion 20–2, *21*
shady conditions 172
shallot 20–2, 62
silverleaf 296
skimmia 406
sloe 80–2, 298, 350
Slow Gardening 129–30
slug pellets 73
slugs and snails 38, 63, 73–5, 78, 104–5, 112, 163, 290, 335, 378, 411, 417
smoke bush (*Cotinus coggygria* 'Royal Purple') 146
snapdragon 286
snow 24–5
snowdrop (*Galanthus*) 13, 14, 15–17, *17*, 29, *40*, 51, 285, 322, 330–2, **333**, 398
Soft Fruit Garden 9, 96, *210*

soft shield fern (*Polystichum setiferum*) 398
soil 8, 261
digging 40
and dry gardens 282–3
lime for 65
temperature 48, 62, 78
waterlogged 219
Solanaceae 334, 380
solidago 321
Solomon's seal 98
sorrel 172, 262
spinach 36, 40, 62, 358
Spring Garden 9, 16, *16*, 31, *40*, 53–4, *54*, 90, 91, 98, 148, 153, 189, 226, 322, 388, 401, *403*, 425
squash 64, 246, 315, 358, 378
Stipa 279, 283, 321, *321*, 360, 360–1
strawberry 60, 210–14, *210*, *213*, *215*, 361
strulch 126–7
sugar-snap pea 108
sunflower 50, 237–9, *237*, 265, 278–9, 279, 313
swallow 343, 415
swede (*Brassica napus*) 65, 407
sweet cicely (*Myrrhis odorata*) 155
sweet pea 132, 218, 239–42, *240*, *285*, 286, 392, 437–8
sweet rocket (*Hesperis matronalis*) 136
sweet violet (*Viola odorata*) 58
sweetcorn 64, 162, 247, 265, 315, 358, 378

T

tellima 93
tender plants 49, 150, 221, 300–2
thistle 201–3, *203*
see also Onopordum
thrush 73, 87, 173, 254, 417, 425
thyme (*Thymus*) 170, 172, 219, 430
tiarella 93
Tilia platyphyllos 8
tithonia (*Tithonia*) 48, 150, 239, 279, 313
toadstools 176
tobacco plant (*Nicotiana sylvestris*) 48–50, 282, 286, 438
tomato 64, 156–61, *156*, 159, *161*, 172, 242, 257–8, 265, 302–6, *302*, 334, 361, 437
tomato blight 159
tomato leaf curl 159
tomato leaf mould 159
tools 437, *438*
Top Veg Garden 9, *19*, *205*, 250, 304, *413*
topiary 324, 351–3, *351*, 398, 422, 424
trachycarpus 406
trefoil 65, 106
tulip 87, 90, 96–8, *96*, *98*, 129, 228, 234, 283, 285, 329, 330, **333**, 368, 398, 420
tulip fire 98
turnip 65, 358, 407
twine 437, *438*

U

umbellifers 153–5, 175, 268, 289, 433

V

varroa mite 216
Venturia inaequalis 373
Verbena bonariensis *240*, 242–4, *244*, 265, 283, 361, 392
vetch 65
viburnum 13, 406
violet 58–60, 83, 239
violet root spot 163
Volutella buxi 327

W

Walled Garden 9, 89, *89*, 91, 93, 130, 132, 136, *136*, 143, 149, *153*, *182*, 184, 185, 188, 189, 222, *223*, 235, 275, 285, *285*, 296, 315, 341, 398
wallflower (*Erysimum*) 48, 129, 135–6, 239, 366–8, *366*, 392
wasp 53, 268
waterlogging 219
weeds 7, 124–7, 226, 283
annual 70, 127
lawn 176
perennial 40, 70, 124–7
white flowers 285–6
white fly 161
wild cherry (*Prunus avium*) 82
wild garlic 83, 260
wildflowers 260, 261–2
willow 86, 218, 226, 389
wind 7, 8, 25, 385–7
windbreaks 25, 385–7
winter aconite (*Eranthis hyemalis*) 14
winter honeysuckle (*Lonicera fragrantissima*) 13, 34, *34*
winter tares 106
wood anemone (*Anemone nermorosa*) 83, 86, *86*
worms 71, 119, 176, 391
Writing Garden 96, *96*, 132, 189, 261–2, 371

Y

yew 8, 9, 83, 310, 351–3, *351*, 406, 419–20, 424
yucca 406

Z

zinnia (*Zinnia*) 24, 48, 150, *150*

ACKNOWLEDGEMENTS

Marsha Arnold has beautifully captured the essence of this garden in sun, wind and rain and has been a calm, wise presence throughout. Everyone at Ebury has been exceptionally helpful but I would especially like to thank Lorna Russell for overseeing the book and Laura Higginson who patiently sifted through my words and pulled them together. My agent Caroline Michel was, as ever, endlessly positive and encouraging. In the garden, Jess Evans and Julia Cooper work expertly and cheerfully behind the scenes and much of the way that the garden appears in these pictures is down to them. Finally by far and away my biggest thanks is to Sarah. This has never been my garden but always ours. It is a true collaboration – and all the better for that.

1 3 5 7 9 10 8 6 4 2

BBC Books, an imprint of Ebury Publishing, 20 Vauxhall Bridge Road, London SW1V 2SA

BBC Books is part of the Penguin Random House group of companies whose addresses can be found at global.penguinrandomhouse.com

This book is published to accompany the television series entitled Gardeners' World, broadcast on BBC Two.

Executive Producer: Gary Broadhurst

First published by BBC Books in 2012. This edition published by BBC Books in 2021.

www.penguin.co.uk

A CIP catalogue record for this book is available from the British Library

ISBN 9781785947827

Printed and bound in Italy by Graphicom s.r.l

Penguin Random House is committed to a sustainable future for our business, our readers and our planet. This book is made from Forest Stewardship Council® certified paper.

Commissioning editor: Lorna Russell
Copyeditor: Helen Griffin
Proofreader: Lesley Riley
Maps: Gardeners' World magazine
Photographer: Marsha Arnold